Trends and opportunities in the production, processing and consumption of staple food crops in Kenya

Trends and opportunities in the production, processing and consumption of staple food crops in Kenya

Edited by
Calvin Onyango
Guenter Unbehend
Meinolf Lindhauer

Bibliographic information published by the Deutsche Nationalbibliothek
The Deutsche Nationalbibliothek lists this publication in the Deutsche Nationalbibliografie; detailed bibliographic data are available in the Internet at http://dnb.d-nb.de.

ISBN 978-3-944331-17-1

Verlag der Wissenschaften GmbH
Bergstr. 70 | D-01069 Dresden
Tel.: 0351/47 96 97 20 | Fax: 0351/47 96 08 19
http://www.tudpress.de

Typesetting: Editors.
Cover: Nicole Hillebrand, Karlsruhe/Germany.
Printed in Germany.

Table of contents

Review papers

Research papers

Preface

Agriculture is the most important economic activity in Kenya, directly contributing 25% to the country's gross domestic product (GDP), and another 27% indirectly. More than one-third of the country's agricultural produce is exported, which accounts for about 65% of Kenya's total exports and 60% of her export earnings. Agriculture is by far the major source on employment in the country. It supports 18% of people in formal employment and 60% of people in informal employment in the country.

The agricultural sector in Kenya comprises four major sub-sectors. These are: industrial crops (tea, coffee, sugar cane, tobacco, sisal and barley), horticultural and associated crops (flowers, fruits, vegetables, nuts and spices), staple food crops (cereals, root and tuber crops, plantains and legumes) and livestock and fisheries sub-sectors. Horticulture and industrial crops account for over 90% of agricultural exports and 50% of agricultural GDP. The livestock and fisheries sub-sector contribute 14% and 6% of agriculture GDP and exports, respectively. The staple food crops (i.e. a food crop that is eaten regularly and in such quantities as to constitute the dominant part of the diet and supply a major proportion of energy and nutrient needs) sub-sector constitute one-third of the agricultural GDP. Practically the entire staple food crop production (99.5%) in the country is consumed internally and only 0.5% is left for export.

Maize is the premier staple food crop in the country. It constitutes 80% of the national cereal production and is associated with national food security. Wheat and rice are the next most important staple food crops in Kenya. However, the demand for maize, wheat and rice in the country commonly outstrips supply due to several technical, institutional and economic factors. The supply deficit is usually met by imports from the regional or international market. Several other staple food crops (i.e. sorghum, millets, cassava, potatoes, sweet potatoes and plantains) are grown across a wide range of agro-ecological zones of the country but do not contribute significantly to the energy and nutrient needs of the Kenyan population. Their production and processing has not been fully exploited, although it is acknowledged that they can substantially contribute to increased farm incomes, improved food security and development of agro-processing industries.

National trends in the consumption of staple food crops are not static. Maize was the staple food crop in Kenya in the 1960s but it now (2013) has the same dietary significance as rice and wheat. Consumers' desire for certain food products is influenced by urbanization, amount of disposable income, and the opportunity cost of labour among other factors. For instance, in Kenya, demand for wheat and rice increases whereas that for maize, sorghum, millet, cassava and sweet potatoes decreases with increasing income largely because the former are more readily available as convenient ready-to-eat foods that require minimal processing. The convenience of these foods is derived from increased mechanization of food processing, which is made possible by the supply of raw materials that meets demand for specific end-use quality requirements. It is easier for food processors to acquire rice or wheat with minimal variability between batches and over time than sorghum, millet, cassava or sweet potato. Thus, consumption of a wide range of staple food crops is best assured when producers meet the end-user requirements of processors by supplying raw materials with defined and consistent quality traits.

The objective of the national conference on *Trends and opportunities in the production, processing and consumption of staple food crops in Kenya* was to communicate the findings of a two-year research project (June 2011 – June 2013) to academic peers, industrialists, policy-makers and regulatory officials. The aim of the project was to investigate the potential of utilizing sorghum and cassava to make gluten-free bread (i.e. bread from non-wheat flours) and composite bread (i.e. bread from wheat and non-wheat flours). Bread was selected as the vehicle to promote these crops because it is a ubiquitous food product that is eaten across all social classes, is culturally accepted across the country and is convenient to handle, store and consume. Some results of this project are presented in Chapter 1 and 3.

The project was undertaken by the Food Technology Division at Kenya Industrial Research and Development Institute (KIRDI, Nairobi, Kenya), and the Department of Safety and Quality of Cereals at Max Rubner-Institut (MRI, Detmold, Germany). The project (Project No.: KEN 10/001) and conference were supported by Bundesministerium für Bildung und Forschung / Federal German Ministry of Education and Research (BMBF, Germany) and Ministry of Industrilaization and Enterprise Development, Kenya. Additional financial support was provided by Deutscher Akademischer Austausch Dienst (DAAD) to enable DAAD-Alumni attend the conference.

We were aware that several other research activities, technology transfer initiatives and policy recalibration efforts are ongoing in the country to develop a vibrant staple food crops sector. It is for this reason that we invited contributions from other stakeholders. The contributions addressed policy and regulation, research, extension, production, processing and consumption of staple food crops in Kenya. Full-length articles of some of the presentations made at the conference are presented in this book. We thank all those who committed their time and effort to contribute to these proceedings, and hope that you, the reader, will be able to identify opportunities for research and technology transfer so as to be able to stregthen the value chain of these crops.

A conference is never complete without delegates who come to listen, question and formally and informally share their knowledge, ideas and experiences with their peers. We are grateful to all those delegates who spared some time from their schedules to participate in this conference.

Dr.-Ing. Calvin Onyango, Kenya Industrial Research and Development Institute
Dipl.-Ing. Guenter Unbehend, Max Rubner-Institut
Prof. Dr. Meinolf G. Lindhauer, Max Rubner-Institut

Acknowledgements

The publication of this book has been made possible with the financial support of Bundesministerium fuer Bildung und Forschung (BMBF) / Federal German Ministry of Education and Research, Germany.

We want to express our gratitude and appreciation to Frau Dr. Sara Kranz and Frau Margitta Rahnert for their assistance in ensuring that the project objectives and budget were strictly adhered to. We also acknowledge the assistance of Ms Anne Wacuka Mutahi, Ms Eunice Akello Mewa and Ms Florence Akinyi, who participated in the project as research scientists. We thank Mr Joseph Nyakoni, Mr Arthur Oyoo, Ms Milcah Mariga, Ms Sara Gacii, Ms Naanyu Karbolo, Mr Benard Endire, Mr Ronald Barongo, Ms Hannah Mugure Kamano-Kamau and Mr Kennedy Olweny Ocholla who tirelessly worked to make the conference a success.

Dr.-Ing. Calvin Onyango, Kenya Industrial Research and Development Institute
Dipl.-Ing. Guenter Unbehend, Max Rubner-Institut
Prof. Dr. Meinolf G. Lindhauer, Max Rubner-Institut

1. Production and consumption of staple food crops in Kenya: 1990 - 2011

Calvin Onyango[1*] and Florence Akinyi[1]
[1]Food Technology Division, Kenya Industrial Research and Development Institute, Nairobi, Kenya.
**Corresponding author: calonyango@yahoo.com*

INTRODUCTION

A staple food is one that is eaten regularly and in such quantities as to constitute the dominant part of the diet and supply a major proportion of energy and nutrient needs. Typically, staple foods are well adapted to the growth conditions in their source areas. Most people live on a diet based on one or more of the following staples: cereals (rice, wheat, maize, sorghum and millets); roots and tubers (potatoes, cassava, sweet potatoes, yams and taro); plantains; legumes (beans, peas and lentils); and animal products such as meat, milk, eggs, cheese and fish. Rice, maize and wheat provide over two-thirds of the world's food energy intake, whereas roots, tubers and plantains are important staple crops for over 1 billion people in the developing world (FAO, 1995). Legumes and animal products are complementary staple foods that play an essential role in meeting nutrient requirements. Cereals, roots, tubers and plantains constitute over two-thirds of the staple foods in the average sub-Saharan African diet, including Kenya, in terms of energy (FAO, 1995). This purpose of this article is to review the production and consumption of cereals, root and tuber crops and plantains in Kenya between 1990 and 2011.

CEREAL CROPS

Cereals are the edible seeds or grains of the grass family, Poaceae (formerly Gramineae). Cereals are the most important food crops globally and occupy over half the harvested area in the world (FAO, 2012). Relative to other staple food crops, cereals are cheap to produce and easy to store and transport. The most important edible cereals worldwide are rice, wheat, rye, barley, oats, millets, sorghum and maize. These grains are a major source of nutrients for humans and animals and are also important industrial raw materials for the production of starch, alcoholic beverages, oil, food sweeteners and bio-fuels. In developed countries up to 70% of the cereal harvest is used as animal feed whereas in developing countries the crop is primarily reserved for human nutrition (Hammes et al., 2005). Cereals and cereal products are an important source of macronutrients including carbohydrates, proteins and fibre; micronutrients such as vitamin E and the B vitamins; and minerals, such as iron, magnesium and zinc. Regular consumption of whole cereal grains may reduce the risk of obesity or weight gain and prevent chronic diseases such as coronary heart disease, diabetes and cancer.

Maize is the most important cereal and staple food crop in Kenya. It accounts for more cultivated area than any other crop and is the most common crop grown by rural poor households for food. It is grown in almost all the agro-ecological zones of the country including regions where it should not be the dominant agricultural crop. It is the main diet for a majority of Kenyans. National and household food security status in the country is synonymous with national maize reserves. The significance of maize as a household food security crop is reflected in the fact that farmers are reluctant to diversify into other crops without assurance of sufficient maize.

Wheat and rice are the second and third most important cereal foods in Kenya. Although maize, wheat and rice constitute the largest share of staple food crop expenditures in the country, their production lags behind the population growth rate due to various technical, economic and institutional constraints. Consequently, the gap between production and demand is constantly increasing thereby necessitating increased imports. Import quantities of these cereals is bound to increase with increasing population, decreasing arable land, changing land-use patterns, land degradation and sub-division and rural-to-urban demographic shifts. The average import quantities of maize, wheat and rice between 1990 and 2010 were 280,000, 500,000 and 130,000 tons, respectively (Figure 1). The corresponding average cost of these imports was US$ 70, 100 million and 40 million, respectively (Figure 2). The total cost of importing these grains was US$ 100 million in 1990 and peaked at US$ 700 million in 2009 before declining to US$ 390 million in 2010. The average wheat and rice imports between 1990 and 2010 represented 60% of total national demand whereas maize imports represented 10%.

Sorghum and millets are drought-resistant crops that are grown in semi-arid agricultural zones of Kenya where maize, wheat and rice cannot consistently produce a reasonable harvest. Although these cereals give relatively high yields per hour of labour spent and are relatively free from major disease epidemics, acreage devoted to them has been declining over the years due to farmers' preference for maize as the major food security crop. Maize production in Kenya exceeds sorghum and millet production by a factor of 25 and 45, respectively. There is no large-scale sorghum or millet farming in the country. These crops are mainly grown in smallholder farms as subsistence crops. Consumption of sorghum and millets is largely confined to their areas of production. Nontheless, consumer interest in these grains is growing because of increasing knowledge of their health benefits. There is evidence to suggest that regular consumption of whole grain cereals could reduce the risk of weight gain or obesity, cardiovascular diseases, diabetes and colorectal cancer (McKevith, 2004; Jones and Engleson, 2009).

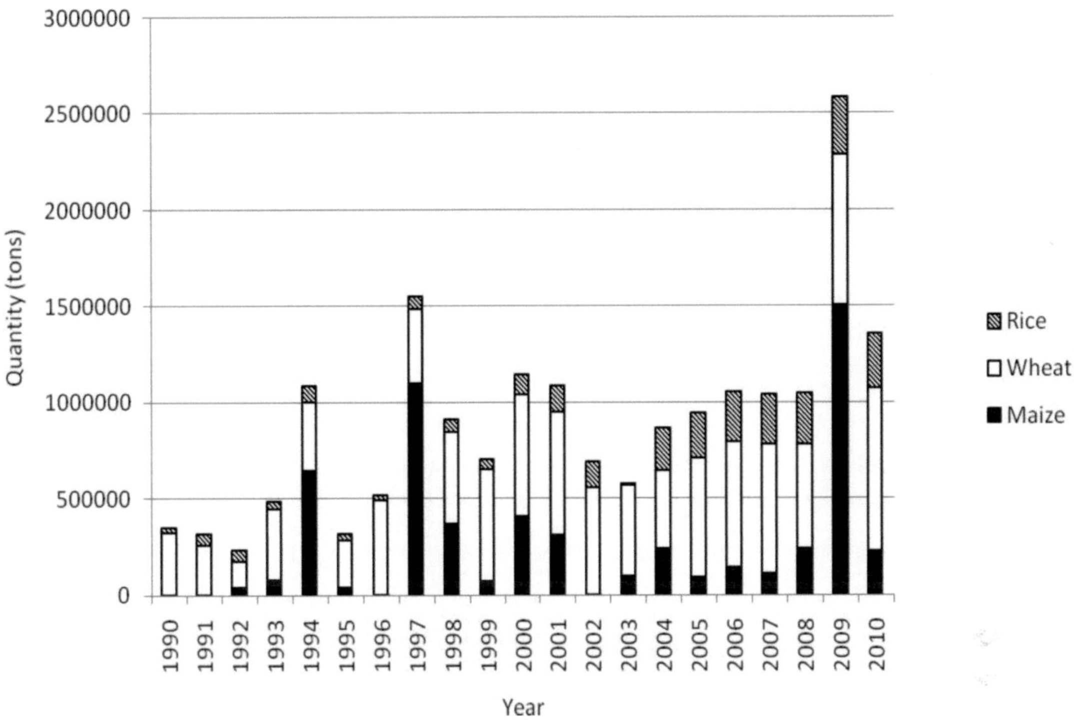

Figure 1. Import quantities (tons) of maize, wheat and rice between 1990 and 2010. Source: FAO (2011).

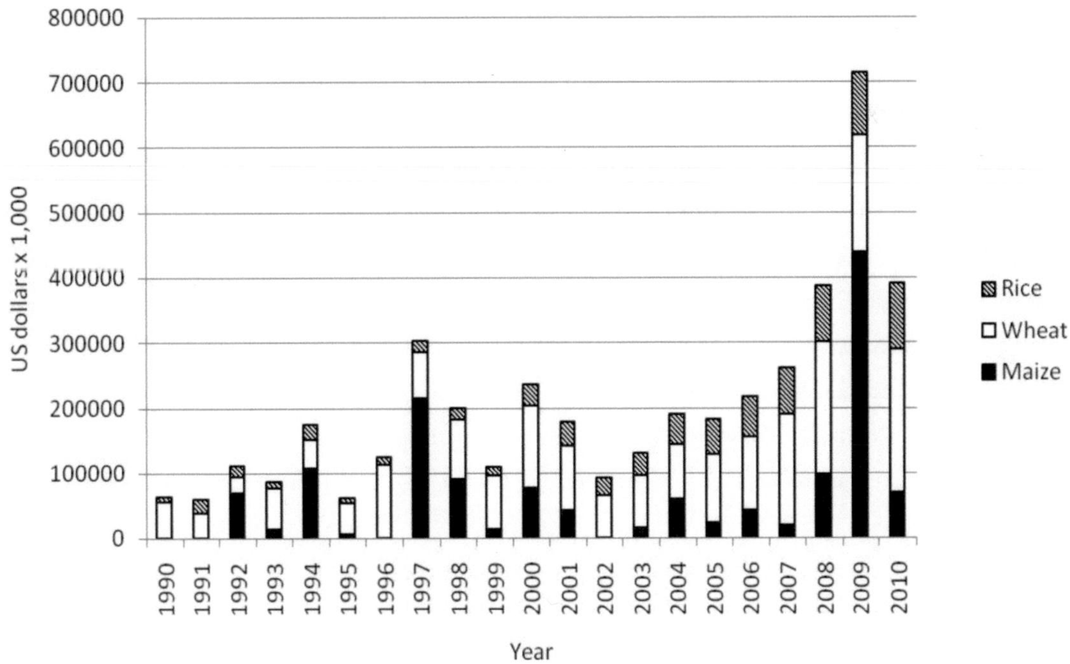

Figure 2. Import cost (US$ x 1000) of maize, wheat and rice between 1990 and 2010. Source: FAO (2011).

Maize

White dent maize (*Zea mays indentata*) is Kenya's main staple food and has dominated the national diet since the beginning of the 20th Century. Maize production increased rapidly from 1963 to early 1980s as the government sought to ensure food self-sufficiency. It then began to decelerate in 1980's due to decreased state intervention. Consequently, within 30 years of independence the country transited from being a net exporter to a net importer of maize. Nonetheless, maize remains important to national and household food security. The government is fully committed to support the production, processing and storage of maize as a strategic grain reserve. Maize is grown by 98% of rural farm households and makes a large share of the households' crop income (FTF, 2011). It accounts for 65% of total staple food caloric intake and 36% of total food caloric intake (FAO, 2011). One-third of the arable land in Kenya is devoted to maize cultivation, which constitutes 80% of the national cereal production (FAO, 2011; Gitu, 2006).

There are about 3.5 million small-scale maize farmers (owning less than 2 hectares) who constitute over 97% of all maize producers in the country and account for more than two-thirds of total annual production from approximately 80% of total annual area planted with maize (ACDI-VOCA, 2010; Odame et al., 2008; USAID, 2010). The remaining one-third of the crop is produced by medium and large-scale farmers on 20% of total annual planted area. Most small-scale farmers use manual labour to produce the crop; intercrop maize with other crops, mainly legumes; and produce maize for household food consumption (30-50%) and sell the rest to earn income. The medium- to large-scale producers depend on maize as a major source of income and sell practically all their harvested produce except for very small quantities (less than 1%) which are consumed at home or used as animal feed and seed (USAID, 2010).

The cultivated area under maize is about 1.9 million hectares (MoA, 2010a) and the potential for further expansion is limited largely by competition on land use from other high-value crops (especially tea, coffee and horticultural produce), sub-division of farms into small uneconomical units, and changing land-use patterns. Trans Nzoia, Uasin Gishu, Narok, Nakuru, Bungoma, Nandi and Bomet are the major maize producing counties in Kenya and produce almost 80% of the crop from about 337,440 hectares (PREMU, 2011). The average maize yield ranges between 1-2 tons/hectare but potential exists to increase it to 6-8 tons/hectare (MoA, 2010a; ACDI-VOCA, 2010). The low yields are caused by declining soil fertility; low utilization of fertilizers, approved seed varieties, and pesticides which, in turn, are attributed to high cost of credit; poor agricultural practices and emerging diseases (Box 1). Generally, maize yields in large-scale production systems are higher than in small-scale systems because of greater use of fertilizers and agrochemicals and higher levels of on-farm mechanization. There is tremendous potential for further increases in maize yields in both small- and large-scale production systems through strategies that include sustained adoption of high yielding varieties, optimal use of fertilizers, improved seed quality assurance and intensification of research on high yielding and drought resistant maize varieties (Gitu, 2006).

BOX 1: MAIZE LETHAL NECROSIS

Diseases have destroyed 64,000 hectares of land under maize…The most affected region is in Rift Valley region where 43,000 hectares have been destroyed…The government had predicted that farmers would harvest 29.9 million bags of maize (*ca. 2.7 million tons*) in 2012 but this has now been revised to 28.9 million bags (*ca. 2.6 million tons*).

Source: The Standard, Thursday, September 13, 2012, Page 12.

Annual maize production between 1990 and 2011 showed extreme year-on-year variation between 2 and 3 million tons/year (Figure 3). The average maize production during this period was 2.6 million tons. In the period under review production exceeded 3 million tons only in 1994, 2006, 2010 and 2011. A good harvest year can suddenly be succeeded by a bad year, which may be a result of factors such as drought, political instability or pests and diseases. Postharvest losses (pest infestation, spillage, spoilage, aging and fungal contamination), which account for 9% of annual production (USAID, 2010), further decrease the amount available for consumption. On the demand side, the national per capita consumption of maize is about 88 kg/person/year (FAO, 2011), which makes an annual demand of about 4 million tons (based on a population of 43 million in 2012). The annual compound growth rate of national consumption between 2004 and 2008 was estimated at 3.7% per year while production declined at about 1% per year (USAID, 2010). If this trend continues, then the country will soon experience maize deficits even in good harvest years. The shortfall in maize production is bridged by imports from South Africa, USA, Brazil, Argentina and the neighbouring countries that make up Common Market of Eastern and Southern Africa (COMESA). There is also substantial undocumented cross-border trade in maize between Kenya and Uganda, which mainly meets the demands of food deficit regions along the common border.

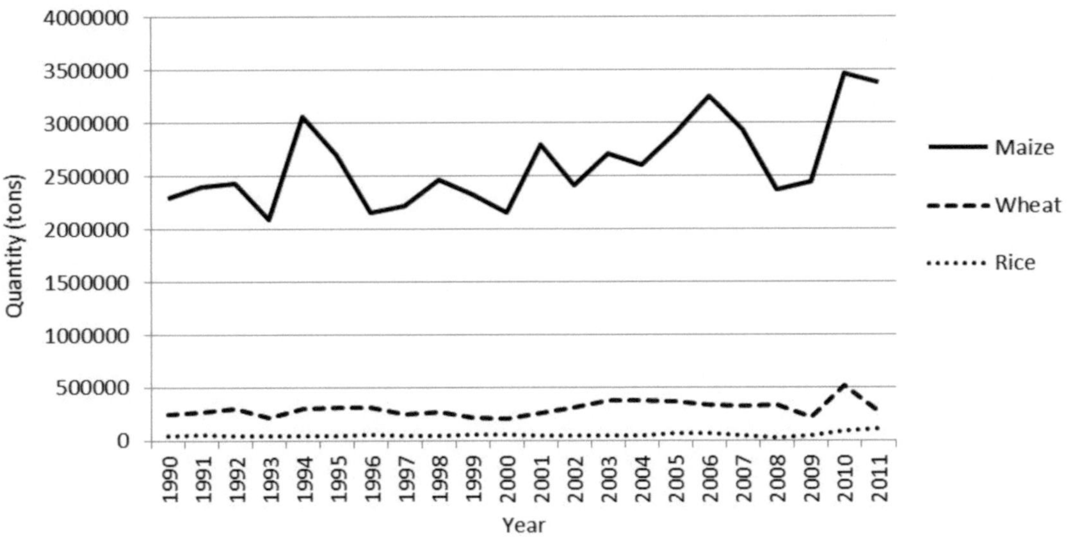

Figure 3. Production (tons) of maize, wheat and rice between 1990 and 2011. Source: FAO (2011).

Household contamination of maize with aflatoxin is a major food safety issue in Kenya. Aflatoxins are highly carcinogenic toxic secondary metabolites produced by various species of *Aspergillus* fungi. Acute exposure to high levels of aflatoxins leads to aflatoxicoses, which can result in rapid death due to liver failure. Chronic exposure to aflatoxins can cause immunosuppression, stunting and kwashiorkor in children, liver cancer (especially in people with hepatitis B or C) and liver disease. Incidences of aflatoxicoses resulting from consumption of contaminated maize frequently occur in eastern and central Kenya (Embu, Kitui, Machakos, Mbeere, Mwingi and Thika). The worst known outbreak of aflatoxicoses in the country took place in 2004 when 125 deaths occurred from 317 reported cases (IFPRI, 2010). In 2010 most farmers in eastern and coastal regions of Kenya could not market their maize at good prices because the aflatoxin levels were several times above the tolerable levels (10 µg/kg or 10 ppb). The outbreaks of aflatoxin poisoning in Kenya usually coincide with periods of nationwide or localized food shortage occasioned by drought or short rains that occur after protracted periods of drought.

Contamination of maize with *Aspergillus* fungi occurs during crop growth and when the produce comes into contact with soil during harvesting, threshing and drying. Storage of inadequately dried grain in poorly constructed and maintained household cribs is also major source of contamination. Although the presence of aflatoxins in food products cannot be completely eliminated, it can be controlled to avoid causing harm to humans. Treating soils with lime, good farming practices (such as timely planting, weeding and pest and disease control), planting the right maize varieties and harvesting at the right time can reduce plant susceptibility to aflatoxins. Adequate drying of the grain in a clean environment with no contact with the soil, storage under cool air temperature and low humidity, removing potentially contaminated kernels and chemical preservation of the grains using ash, acetylic or phostoxin in storage units can prevent pests and reduce the likelihood of contamination (IFPRI, 2010).

Although maize is the most important staple food crop in the country, Kenya's maize policy is a major contributor to the recurrent national food crises. Kenya's staple food sector is completely liberalized, with the exception of maize whose prices are occasionally set by Government thereby influencing free market forces, primarily because National Cereals and Produce Board (NCPB – a government institution that is mandated to purchase, store, market and generally manage cereals grains and other produce) is a major player (USAID, 2010). Farmers hardly ever make decent profits whereas consumers are forced to purchase maize products at prices above world market prices. Maize prices in Kenya are among the highest in sub-Saharan Africa. In 2011 Kenyans paid US$ 45 per bag (i.e. 90 kg) of maize, which was 70% above the world market prices (PREMU, 2011). The high price policy maintained by NCPB and the imposition of numerous non-tariff barriers means that Kenyans have to pay higher prices than they would under a market-based system (PREMU, 2011; Jayne et al., 2005). The main groups affected by Kenya's maize policy and the resultant price fluctuations are the non-agriculturally productive households (a category that includes most of the urban poor and landless), households that produce food for their own consumption but also purchase additional food (about 60% of Kenyan farmers) and households that are net sellers of maize (less than 2% of Kenya's farmers, PREMU, 2011). The World Bank has stated that Kenya can assure all-year round availability of competitively priced maize by separating the three functions of NCPB (which are famine relief, maintenance of commercial stocks and strategic grain reserves), reducing import duties and liberalizing trade within the East African Community (PREMU, 2011).

About 86% of the maize produced in Kenya is dry milled for human consumption. The rest of the production is distributed as follows: postharvest losses, 9%; animal feeds, 3%; seed, 1%; industrial uses 1%. Maize for human consumption is commonly dry milled into maize meal, flour or grits. Maize flour is sold to consumers who use it to make thick porridge (ugali) or thin porridge (uji), whereas maize meal and grits are sold to breakfast cereals manufacturers and brewing industry. Maize millers in the country can be grouped into three categories based on their milling capacities. The first group comprises about 109 medium- to large-scale industrial maize (and/or wheat) millers who are members of Cereal Millers Association (CMA). The second group comprises about 100 small-scale maize millers most of whom are affiliated to the United Grain Millers and Farmers Association (UGMFA). The third group comprises a large but unknown number of micro-millers (about 10,000) commonly known as *posho* millers (i.e. hammer mills). In addition, NCPB sometimes contracts millers to carry out maize processing on their behalf (USAID, 2010). The CMA estimates the national maize milling capacity at 1.62 million tons/year, out of which 85-90% is milled by 23 large-scale millers (Figure 4). The posho millers have a combined milling capacity of 10-15% of total national installed milling capacity (USAID, 2010). The current average capacity utilization for CMA members is about 55% of total installed capacity. The underutilization of capacity is due to insufficient supplies of maize (and wheat) and competition arising from increased capacity in maize (and wheat milling) in the neighbouring countries (USAID, 2010).

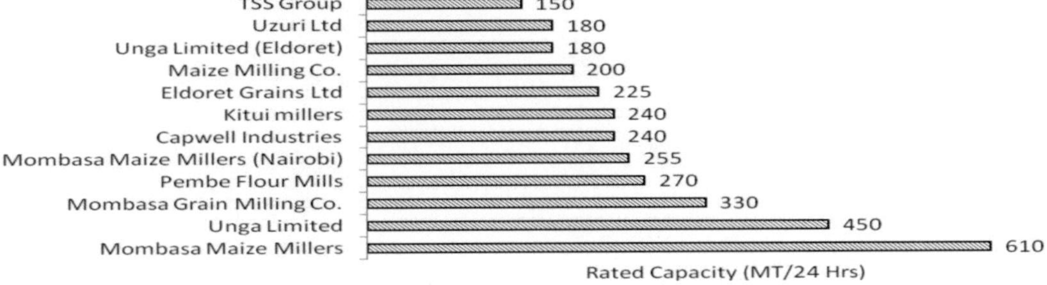

Figure 4. Milling capacities of 23 largest maize milling firms in Kenya (Source: USAID, 2010).

Wheat

Wheat (*Triticum aestivum*) is the second most important staple food nationally after maize. Its consumption accounts for 17% of total staple food caloric intake in Kenya (FAO, 2011). It is grown in the cooler and medium-rainfall counties of Nakuru, Uasin Gishu, Trans Nzoia and Narok at about 1,800 meters above-sea-level. It is mainly grown by small-scale farmers (93% of wheat farmers) whose output is 25%. The output of large-scale wheat farmers (0.3% of wheat farmers) is 45% whereas that of medium-scale farmers (6.6% of wheat farmers) is 30% (Table 1). The cultivated area under wheat is about 154,000 hectares (Dixon et al., 2010) but this can be increased to 500,000 hectares primarily in Narok county (Makanda and Oehmke, 1993). Wheat yields vary greatly between small-scale and large-scale farmers (Table 1). The average wheat yield in the country is about 2.4 tons/hectare, which is higher than the average yield in Eastern and Southern Africa (1.8 tons/hectare) and North Africa (2.3 tons/hectare). Zambia and Zimbabwe are the only African countries with higher wheat yields than Kenya at 4.7 and 3.9 tons/hectare, respectively (Dixon et al., 2009).

Wheat production in Kenya shows substantial year-to-year fluctuation. Between 1990 and 2011, production fluctuated between 200,000 and 400,000 tons/year. It peaked at 513,056 tons in 2010 before declining to 268,482 tons in 2011 (Figure 3). Wheat production has steadily deteriorated from self-sufficiency in the 1960s to an import level of 60% of the national demand. Wheat is now the single largest food import in Kenya and constitutes a major portion of emergency food aid. Most of the imported wheat is from Argentina, USA, Ukraine and Russia. Wheat imports ranged between 130,000 and 840,000 tons/year (Figure 1) at an annual cost of US$ 23-220 million between 1990 and 2010 (Figure 2). The annual cost of wheat imports in the last five years covered by this review (2006-2010) has ranged between US$ 170-220 million. The future demand for wheat is bound to increase with increasing urbanization and population growth and changing consumption habits (as incomes increase, demand for wheat increases as that for maize declines). Consumption of wheat has also been encouraged by national policies that have helped to make imported wheat more available and affordable than domestically produced cereals. The national per capita consumption of wheat is currently about 26 kg/person/year (FAO, 2011) making an overall demand of 1 million tons/year (based on a population of 43 million in 2012). This demand can only be fully met if the cultivated area is doubled from the current 154,000 hectares to 300,000 hectares and the yield increased to 3 tons/hectare.

Due to various technical, economic and institutional factors, the producer price of wheat in Kenya is always higher than world prices. Wheat production in the country is constrained by several factors, which include high post-harvest losses, inappropriate production technologies, inadequate access to credit and extension services, poor crop management practices (such as low fertilizer application and use of non-certified seeds), soil acidity and reliance on rain-fed farming. Research done by Tegemeo Institute of Agricultural Policy and Development has shown that imported wheat (exclusive of duty) is cheaper than locally produced wheat (Table 1). Wheat production is most efficient in large-scale farms and least efficient in small-scale farms. It can therefore be argued that wheat productivity can be enhanced if the government placed more emphasis on technical improvement in large farms with adaptive research for transmitting these improvements to small-holders. The government could also boost wheat production by engaging more in the provision of research, extension, credit, marketing and storage services to wheat farmers. A predictable policy on wheat imports can also help producers make long-term investment decisions. The current scenario whereby import duties and levies are constantly revised is not conducive to the development of this sector.

Table 1. Comparison of the costs of local versus imported wheat.

Production costs (efficiency categories)	Most efficient	Average	Least efficient
Farm size category*	Large scale	Medium scale	Small scale
Yield (tons/hectare)	3.6	2.0	1.3
Number of farmers (% of total in brackets)	30 (0.3%)	600 (6.6%)	8,400 (93.1%)
Production, tons (% of total in bracket)	257,000 (45%)	164,250 (30%)	136,688 (25%)
Total cost Nairobi (US$/ton)	263	287	336
Imports from Argentina			
Import price categories	Minimum	Average	Maximum
CiF** to Mombasa (US$/ton)	150	175	200
Total cost in Nairobi inclusive 35% duty (US$/ton)	303	341	378
Total cost in Nairobi exclusive 35% duty (US$/ton)	246	274	302

Adopted with modifications from Nyoro, J. (2006).
*According to Cereal Growers Association large-scale farmers own all their farm machinery and implements and cultivate 100 or more acres, medium-scale farmers may own some machinery and cultivate from 20-100 acres whereas small-scale farmers cultivate less than 20 acres and mainly hire machinery.
**CiF: cost, insurance and freight.

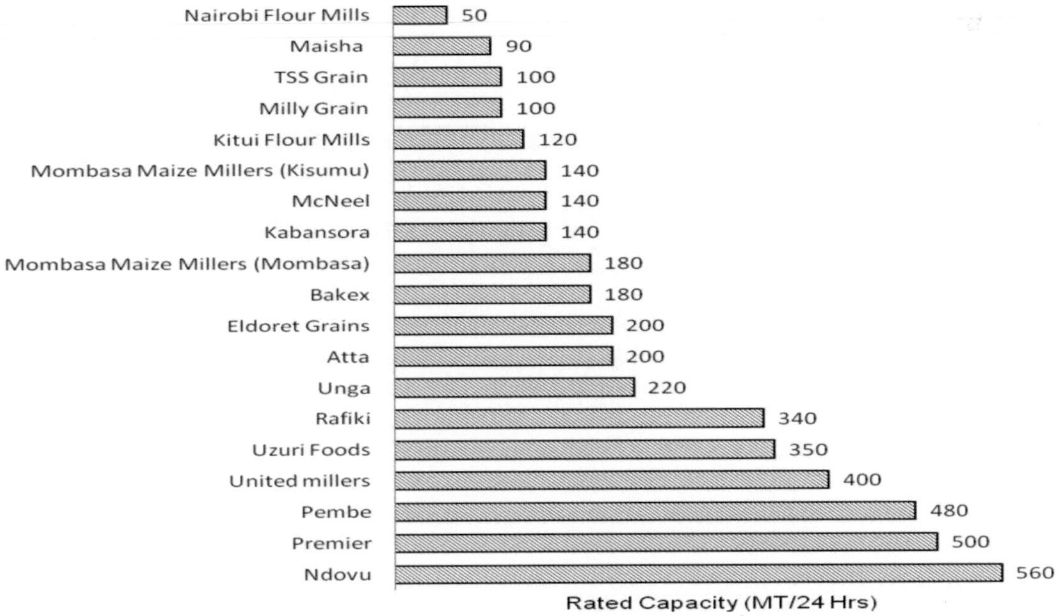

Figure 5. Milling capacities of 18 largest wheat milling firms in Kenya (Source: USAID, 2010).

Value addition to wheat in the country is limited to milling it into flour. Wheat farmers sell their wheat directly to millers or NCPB who store it in silos for later sale to the millers. Kenya has 162 wheat grain millers with highly diverse production capacities. The 18 largest millers (Figure 5) have a combined installed capacity of 1.5 million tons/year but utilize only 50% of the installed capacity (USAID, 2010). The major constraints affecting these millers are limited availability of wheat, poor quality of wheat, high domestic prices, high transport costs and unpredictable government regulations. The milled wheat is sold to bakeries and households via retail shops and supermarkets. The 148 registered bakeries in the country produce 93,000 tons of bread annually and 91,000 tons of biscuits and other products. The by-products of the local wheat milling industry are bran and pollard, which are sold to local animal feed industries.

Rice

Rice is the third most important cereal crop in Kenya after maize and wheat. About 300,000 farmers earn their livelihood from rice production (MoA, 2008). Rice is grown mainly by small-holder farmers as a commercial and food crop in designated irrigation schemes that are managed by National Irrigation Board. A small number of small-holder farmers also grow the crop under rain-fed conditions. About 80% of the rice grown in Kenya is from irrigation schemes while the remaining 20% is produced under rain-fed conditions (MoA, 2008). The major rice-growing regions in the country are in Mwea, Kano, Bunyala, Ahero and the Tana delta basin. The Mwea Irrigation Scheme accounts for over 75% of the rice produced in Kenya (MoA, 2010a) and 97% of all irrigation schemes (USAID, 2010). Rain-fed rice is grown in Migori, Busia and Kuria in western Kenya; Tana Delta and Msambweni in eastern Kenya; and Kwale and Kilifi counties in the coast region. The Bura and Pekerra irrigation schemes also have tremendous potential for rice production.

Kenya has an estimated irrigation potential of 1.3 million hectares out of which 114,600 hectares (ca. 8%) has been developed (GoK, 2010), and 22,000 hectares (ca. 2%) is cultivated with rice (MoA, 2010a). Most of the unutilized irrigation potential lies in the Lake Victoria basin. Previously the National Irrigation Board had a large stake in irrigated rice production in the country through provision of extension services, inputs and post harvest processing and marketing facilities. But with the liberalization of the market in 1990s, the National Cereals and Produce Board (NCPB), regional bodies (such as Lake Basin Development Authority and Tana and Athi River Development Authority) and private organizations (such as Mwea Farmers Multipurpose Cooperative Society, Capwell Industries and Dominion Farms) have increased their presence in the rice value chain.

Rice production fluctuated between 40,000 and 55,000 tons between 1990 and 2005. It peaked at 64,840 tons in 2006 before declining to 47,256 tons in 2007 and further to 21,881 tons in 2008. Thereafter it almost doubled to 42,202 tons in 2009 and further to 80,042 tons in 2010 (Figure 3). Production in 2011 was 111,229 tons. The average rice yield is estimated at 2 tons/hectarc (MoA, 2010a), which is significantly below the genetic potential of 10 tons/hectare that can be realized from 2-3 crops planted in a year. The differences in yields are starkly evident when private firms are contrasted with government-managed firms. Dominion Farms Limited, a privately owned entity that grows and mills rice in Yala delta in western Kenya, achieves yields of between 8.6 and 10 tons/hectare, whereas the average rice yields in schemes operated by National Irrigation Board is about 3.7 tons/hecatre (USAID, 2010). Rice yields in the country are constrained by the high cost of production, high cost of farm inputs (labour, fertilizers, agro-chemicals and seeds) loss of genetic vigour (due to use of the same varieties over many years), inadequate extension support, prevalence of water-borne diseases that reduce productivity and declining water distribution efficiency. In response to these constraints, the Ministry of Agriculture, Livestock and Fisheries has developed a National Rice Development Strategy (MoA, 2008), that proposes a raft of measures that should increase rice yields to 5.1 tons/hectare by 2018.

Kenyan farmers grow mainly aromatic white long grain rice variety (also known as basmati or pishori rice). Kenyan consumers prefer this variety because of its superior cooking qualities and aromatic taste in comparison to other local and imported varieties. Rice forms an important part of the diet of most people living in urban areas of Kenya but consumption is still limited in the rural areas. The national per capita consumption of rice is about 8 kg/person/year (WARDA, 2007), which makes an annual demand of about 340,000 tons (based on a population of 43 million in 2012). National rice consumption is increasing at an annual rate of 12% compared to 4% for wheat and 1% for maize (MoA, 2008). If this trend is maintained then rice will soon surpass wheat as the country's second most important grain after maize. Increasing rice consumption is associated with increasing disposable income in urban areas and changing consumer tastes. Rice is also an important staple food in large institutions such as boarding schools, hospitals and hotels because it is relatively easy to prepare.

Because local rice production is unable to keep pace with demand, the country has to import almost 60% of her requirements from Pakistan, Thailand, China, Vietnam and India. Rice imports ranged between 27,000 and 300,000 tons/year between 1990 and 2010 (Figure 1) at a cost of US$ 8-100 million/year (Figure 2). In recognition of the importance of rice in enhancing national food security, the Ministry of Agriculture, Livestock and Fisheries has developed a National Rice Development Strategy (MoA, 2008) which outlines the roadmap to enhance rice research, production, processing, marketing and utilization in the country. The strategy seeks to increase the area planted with rice from the current 22,000 hectares to 35,150 hectares. It also seeks to increase annual rice production by 9.3% to reach 200,000 tons by 2020 and 500,000 tons by 2030 (Figure 6).

Once paddy rice (also referred to as unprocessed rice or rough rice) has been harvested from the field it is milled to separate the kernel from the husk and the bran. The country has seven major rice mills. Mwea

Rice Mills in Mwea, LBDA Rice Mills in Kisumu, Western Kenya Rice Mills, and Tana Delta Rice Mills in Hola are owned and operated by the National Irrigation Board or the government. The three private rice mills are owned and operated by Dominion Farms Limited (in Yala), Capwell Industries (in Thika) and Mwea Farmers Multipurpose Cooperative Society (in Mwea). Additionally, there are over 350 mills in Mwea and a large but undocumented number of small rice mills in the rice-growing areas of Kenya. The installed milling capacities of the rice mills in the country range from 1.5 tons/day for Mwea Farmers Multipurpose Cooperative Society to 24 tons/day for Mwea Rice Mills (Figure 7). The average operating capacity for the rice milling industry in Kenya is about 40% (USAID, 2010). Mwea Rice Mills has the lowest operating capacity at 13.5% whereas Capwell Industries has the highest operating capacity at 85%.

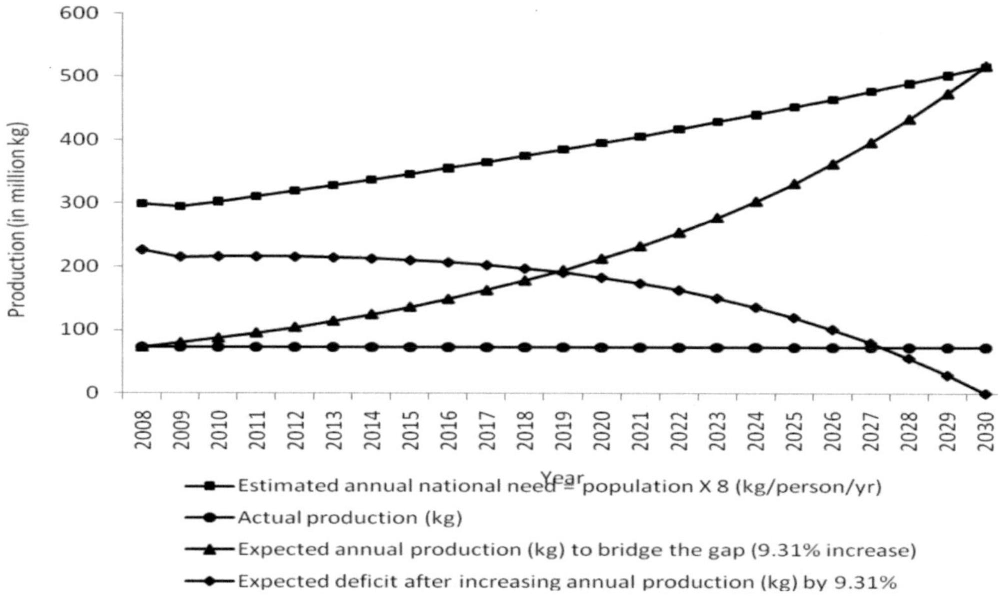

Figure 6. Projections on production and consumption of rice by 2030. (Source: MoA, 2008).

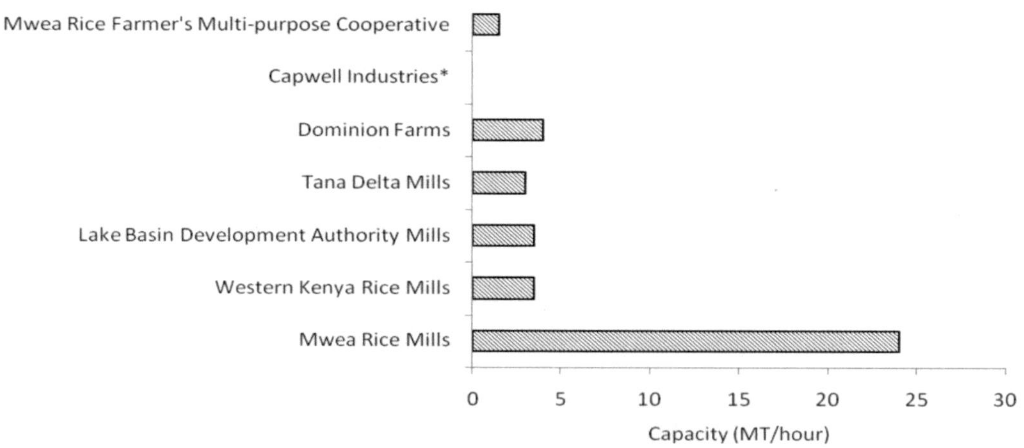

Figure 7. Milling capacities of seven largest rice milling firms in Kenya. (Source: USAID 2010).
*Data not available.

Sorghum and millets

Sorghum (*Sorghum bicolor* (L.) Moench), finger millet (*Eleusine coracana* (L) Gaertn.) and pearl millet (*Pennisetum glaucum* (L) R. Br.) are subsistence crops grown by resource-poor, small-scale farmers as food security alternatives to maize, wheat and rice. In terms of output, the Lake Victoria basin produces about 51% of sorghum crop, whereas the eastern and Rift Valley regions account for 42 and 7%, respectively, of total annual sorghum production. The major regions in the country that grow millets are the eastern region (52%), Lake Victoria basin (32%) and Rift Valley region (15%) (USAID, 2010). Sorghum and millets are versatile crops that outperform maize, wheat and rice in these areas because of their low input and water requirements, greater resistance to diseases and weeds and high production potential per unit of land and per hour of labour spent. However, acreage devoted to these crops has been declining over the years due to greater priority given to maize as the main food security crop (maize production exceeds sorghum and millet 25 and 45 times, respectively) and inefficient production systems (Box 2). Though expansion in area planted is critical to increased productivity, it is not a long-term sustainable solution. Increased production of these grains can be guaranteed by adoption of improved varieties and better crop management.

BOX 2: MECHANIZATION OF SORGHUM AND MILLET PRODUCTION IN KENYA

Sorghum and millet production are labour intensive activities with low levels of mechanization. The level of mechanization as indicated by percent of power for sorghum production from human, animal and tractor sources is 77, 18 and 5%, respectively. Mechanization levels will not increase as long as farm sizes continue to decrease and producers do not meet the end-user requirements of processors.

Source: Wortmann et al.(2006).

Sorghum is cultivated in about 173,000 hectares with mean yields of 0.5 tons/hectare (MoA, 2010a), which is below the genetic potential of 3-5 tons/hectare (KSC, 2012). Between 1990 and 2011, sorghum production fluctuated between 80,000 and 160,000 tons. The highest production for the period under review was 164,066 tons in 2010 whereas the lowest production was 54,316 tons in 2008 (Figure 8). Approximately 60% of the sorghum crop in Kenya is produced in sole crop and the rest is intercropped with maize, millet, cowpea, cassava or common bean (Atlas of Sorghum, 2006). The major improved sorghum varieties in Kenya are KARI Mtama I, Serena, Seredo and Gadam. These varieties have been developed by Kenya Agricultural Research Institute (KARI), and are suitable for the different agro-ecological zones of the country. They mature between 3 and 8 months and have a genetic potential of 2.7-3.3 tons/hectare (USAID, 2010). Farmers choose the cultivars to plant based on certain phenotypic characteristics. Plant height is a major phenotypic characteristic with most farmers preferring plant height of less than 2 m. Brown and red grain types with high tannin content are liked by farmers because they are not preferred by birds and they suffer less quality loss during storage. Cultivars that develop erect position of the panicle are preferred over those that develop a crooked or goose-neck. Threshability is also an important consideration in choice of cultivar to be planted (Wortmann et al., 2006).

Sorghum yield and production in the country is affected by numerous abiotic and biotic constraints. The major plant diseases that affect sorghum production in Kenya are grain mold (which is caused by several fungi), late blight, rust, anthracnose and smut. Stem borer complex, shoot fly, midge, cutworm, army worm and aphids are the most important insect pests. Other important biotic constraints to sorghum production in the country are striga and weeds. Significant quantities of sorghum grain are also lost to domestic and wild animals, and birds, the most important being Quelea. Soil water deficit and soil fertility constraints (i.e. nitrogen, phosphorous and potassium deficiencies) and salinity are the major abiotic constraints to sorghum production in Kenya. Losses of grain sorghum are also occasioned by other abiotic constraints such as inappropriate growth temperature, low radiation, wind, soil crusting and root barriers (Wortmann et al., 2006).

Millet is cultivated in about 104,000 hectares with mean yields of 0.5 tons/hectare (MoA, 2010a). These yields correspond to the yields of improved varieties (i.e. Katumani and P224) that Kenya Agricultural and Research Institute (KARI) has released to Kenya Seed Company for sale (KSC, 2012). During the period under review (1990-2011), millet production never exceeded 120,000 tons/year. The lowest production was recorded in 1996 at 23,159 tons and the highest in 2007 at 119,599 tons (Figure 8). Millet is commonly grown in the same ecosystem as sorghum, and therefore is also affected by many of the biotic and abiotic constraints that affect sorghum production. The main diseases are that affect millet production are leaf blight, blast, smuts (leaf smut, covered kernel smut or long smut), Ergot and head bugs. The main millet pests are birds (such as Quelea), shoot fly, African boll worm and stem borers.

In a paper published in 1993 by KARI/MIAC, it was estimated that sorghum and millet production would have to increase by 300% by 2013 to meet the demand imposed by the high population growth. This means that production in 2012 should be about 300,000 and 180,000 tons for sorghum and millets, respectively. The per capita consumption of sorghum is about 3.8 kg/person/year, whereas for millets it is about 2.2 kg/person/year (USAID, 2010). Thus, for a population of 42 million, sorghum and millet demand in 2012 was about 160,000 and 92,000 tons, respectively. Although sorghum and millet production show extreme year-on-year variation, the average production for these grains was 109,579 and 56,700 tons, respectively, for the period between 1990 and 2011 (Figure 8). It can therefore be concluded that Kenya is not sufficient in these grains.

Figure 8. Production (tons) of sorghum and millet between 1990 and 2011. Source: FAO (2011).

Grain sorghum utilization in the country is distributed as follows: human food, 53.8%; losses 11.3%, animal feed 10%; seed, 1.9%; and processing 23% (USAID, 2010). Most of the sorghum used as human food is milled into whole flour in privately owned *posho* mills (i.e. hammer mills) with milling capacities of 70-200 kg/hour. Sorghum stover is used by resource-poor households as cooking fuel, construction material and mulch. Industrial processing of the grains is slowly becoming more vissible. The grains can be popped and also made into porridge, bread, malts and beer. The grains can be boiled like rice or used as animal feed whereas the stover can be used as grazing silage and hay. East African Malting Limited is promoting the cultivation of Gadam sorghum in the country because it has good malting properties and can be used as raw material in industrial production of lager beer. Sweet sorghum is being promoted for its juicy leaves and stems that are rich in sugars, which can be converted to ethanol for diverse industrial applications. The functional health benefits of whole sorghum and millet grains derive from their slow rate of digestion. Sorghum is also rich in several phytochemicals (i.e. polyphenols, anthocyanins, phytosterols and policosanols), which have positive health properties (Awika and Rooney, 2004).

ROOT AND TUBER CROPS AND PLANTAINS

Root and tuber crops are plants that store edible material in subterranean root, corm or tuber and provide energy in the human diet in the form of carbohydrates. The important harvested part from cassava and sweet potatoes are the roots while from potatoes, arrow roots and yams are the tubers. Plantains are members of the banana family (Musaceae) and are popularly known as cooking banana because they are starchy, low in sugar and must be cooked prior to eating. Roots, tubers and plantains are rich in highly digestible starch and provide more dietary energy per hectare than cereals. However, they are low in protein and other nutrients (Bienvenido, 1999). The high moisture content of these crops means that they are bulky, highly perishable and costly to transport. They have to be processed or stored under modified conditions soon after harvest to minimize post-harvest losses.

Root and tuber crops are the second most important food crops in Kenya after cereals. They are mainly grown in small-holder farms across a wide range of agro-ecological zones of the country and contribute significantly to the food security needs of Kenyans. Potato is an important food and cash crop in the medium and high rainfall areas of Kenya, whereas cassava and sweet potato are important food security crops in the marginal agricultural zones. Plantains are relatively insignificant in Kenya as compared to neighbouring Uganda where they are the staple food. Per capita consumption of plantains is 19 kg/year in Kenya and 140 kg/year in Uganda (FAO, 2011). Yams and arrow roots are insignificant tuber crops in Kenya. They are cultivated in small-holder farms in the central region of the country in about 3,500 hectares of land. Average production between 2005 and 2009 was 4,400 tons for yams and 25,000 tons for arrow roots (MoA, 2010a).

The root and tuber crops sub-sector is faced with a number of challenges which include low and declining productivity, high losses due to pests and diseases, lack of adequate quantities of clean and healthy planting materials, limited and inefficient land use, high cost and inadequate use of farm inputs, inaccessible credit facilities, poor postharvest management practices, low value addition and marketing inefficiencies (MoA, 2010b). In recognition of the importance of root and tuber crops to national food security, the Ministry of Agriculture, Livestock and Fisheries has developed a National Roots and Tuber Crops Policy (MoA, 2010b). This policy has identified the challenges facing this sub-sector and proposed interventions based on the value chain approach which encompass seed production system, research and technology development, extension service delivery, post-harvest handling, marketing and processing, credit and financial services, infrastructure development, stakeholder involvement and institutional arrangements (MoA, 2010b).

Potato

Potato (also known as Irish potato) is a tuber, the tip of an underground stem that swells with stored starch and water. It is the second most important food crop in Kenya after maize (MoA, 2010b). Potato (*Solanum tuberosum*) is an important food and cash crop in the medium and high rainfall areas of Kenya which lie at an altitude of 1500-3000 meters above sea level. These include slopes of Mount Kenya (Meru, Embu and Kirinyaga counties) and parts of Laikipia and Aberdare range (Nyeri, Muranga, Kiambu and Nyandarua counties). It is also grown in the highlands of Mau escarpment (Mau, Narok and Molo), Tinderet, Nandi escarpment and Cherangani hills. Small quantities are also cultivated in Kericho and Kisii counties and around Taita Hills. The high energy content, high market prices and ease of production has made potatoes an important component of urban agriculture.

Kenya has about 790,000 potato farmers who cultivate 108,000 hectares (MoA, 2010b). The average potato yield is about 24 tons/hectare (MoA, 2010b), which compares unfavorably with ideal yields of 40 tons/hecatre (FAO, 2011). Potato yields in the country are constrained by limited use of quality seeds and inputs. Production fluctuated between 600,000 and 2.5 million tons/year between 1990 and 2011 (Figure 9). In the period under review (1990-2011) potato production exceeded 2.5 million tons only in 2005, 2008 and 2010. The lowest production was recorded in 1992 at 632,572 tons. The national per capita consumption of potatoes is about 31 kg/person/year (FAO, 2011), making an overall demand of 1.3 million tons/year (based on a population of 43 million in 2012). Thus, potato is the only staple food crop in the country where production currently exceeds demand.

Potato is mainly consumed in the fresh form in the country. Consumers purchase raw potatoes from open air-markets or supermarkets and prepare it themselves at home. The most common form of preparation is by boiling it with other foods such as legumes, cereals and meat. It is also commonly pounded and eaten as a puree either alone or with dehulled maize and beans. It may also be baked prior to consumption. Consumption of processed potato products is also common in the urban centres. French fries (potato chips or simply chips) is a popular fast-food in the urban centres that is made by deep frying potato strips in oil. Nairobi alone has over 800 restaurants that sell French fries (MoA, 2010b). Industrial level processing of potatoes is mainly in the production of snack foods such as crisps. There exists a huge untapped potential for the industrial production of frozen or dehydrated potatoes. Whole potatoes can be used as livestock feed. Starch can be extracted from potatoes and sold in the native form or converted to several other industrial products. Potato waste water is rich in proteins, non-polymeric nitrogen compounds, sugars and minerals. Proteins can be recovered from the waste water and used in animal feed formulations. The deproteinised water is an ideal medium for various industrial microbial processes. The by-products from potato starch mills can also be used as fertilizer.

Cassava

Cassava (*Manihot esculenta*) is the elongated root of the plant. It is a vital household food security crop in the Lake Victoria basin, coastal belt and the semi-arid regions of eastern Kenya. These regions account for 64%, 20% and 15%, respectively, of the annual cassava crop production in the country (USAID, 2010). Cassava is an invaluable food security crop because it is drought tolerant, has a high production potential per unit of land and high return of food calories per unit energy input in its cultivation. Cassava is a valuable crop for resource-poor farmers because the planting material, in the form of the stem, is always available. About 60% of the crop is grown in pure strands whereas the rest is intercropped with maize, legumes, bananas, sorghum, millet or coconut trees or planted to demarcate sections of the land. The area under the crop declined from 86,190 hectares in 1995 to 51,514 hectares in 2008 (MoA, 2010b) due to neglect, emerging pests and diseases and declining demand. Cassava production in the country is constrained by an underdeveloped market and lack of access to disease tolerant varieties (MoA, 2010b). During the period under review, cassava production fluctuated between 300,000 and 800,000 tons/year (Figure 9). The highest cassava production was attained in 2009 at 819,967 tons. This was followed by 739,983 tons in 1996. The average cassava yield in the country is about 20 tons/hectare against a potential yield of over 50 tons/hectare.

The perception of cassava as a 'poor man's crop' and the fear of cyanide poisoning has greatly affected its status as a staple food crop in the country. Cases of death due to consumption of raw or inappropriately processed roots are frequently reported in the country. It is unarguable that cassava contains cyanogenic glycosides which are converted to toxic hydrocyanic acid when the cells are ruptured. However, the cyanide content of the roots can be reduced to non-toxic levels if they are correctly processed. Hydrocyanic acid is a volatile compound that evaporates readily at temperatures above 28°C and dissolves readily in water. This means that simple procedures such as shredding, pressing, washing, fermentation and drying can reduce the cyanide content to non-toxic levels. Residual amounts that are consumed can be detoxified in the body provided the individual eats a high protein diet that is rich in sulphur-containing amino acids.

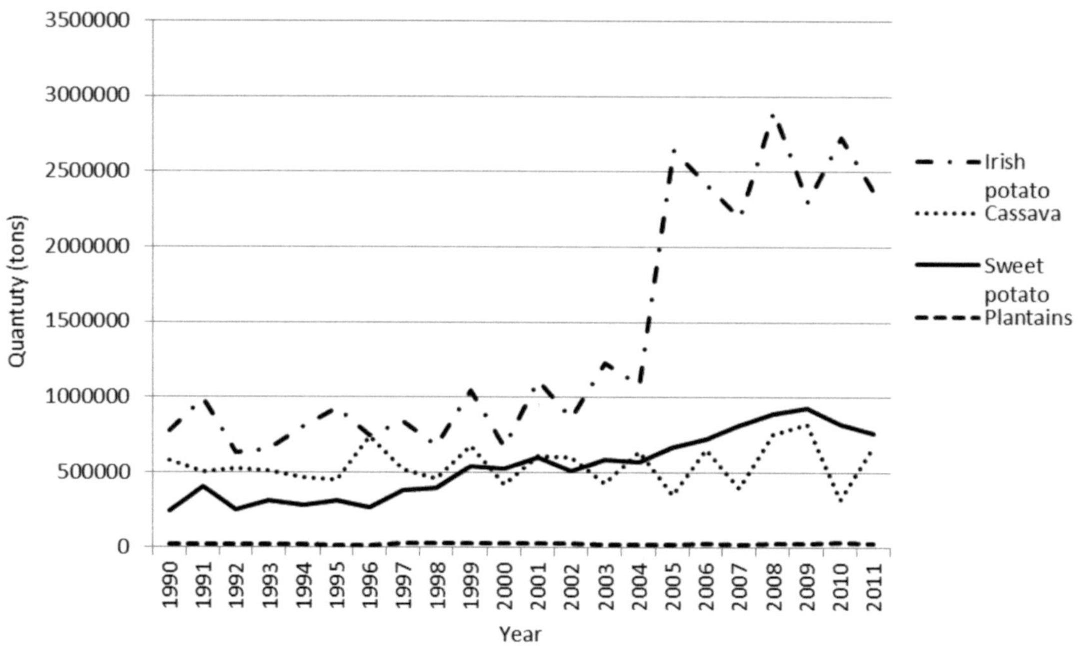

Figure 9. Production (tons) of potato, cassava, sweet potato and plantains between 1990 and 2011. Source: FAO (2011).

The national per capita consumption is about 15 kg/person/year. Thus, for a population of 42 million Kenyans in 2012, cassava demand was about 630,000 tons. Although cassava production shows extreme year-on-year variation, the average production for the period between 1990 and 2011 was 549,887 tons (Figure 9). It can, therefore, be concluded that Kenya is not sufficient in cassava production.

Cassava deteriorates rapidly after harvest. Farmers attempt to overcome this problem by leaving the roots in the ground until when they require them for consumption or sale. The disadvantage here is that a large portion of the land is occupied by the mature crop and is unavailable for other farming activities. Also, the starch content of the root declines and it become fibrous and woody. About 97% of the cassava that is produced in the country is used as food either fresh or chipped, dried and milled into flour. Cassava flour is usually blended with sorghum or millet flours and is used to prepare porridge. Although there are several industrial uses of cassava, these have not yet been exploited in Kenya. The tubers can be dehydrated and pelleted then used as animal feed. Cassava tubers are a rich source of starch which can be used as raw material by the food and non-food industry (Figure 10). Starch can be hydrolysed to glucose and further to high fructose syrup. It can be converted further into ethyl alcohol, acetone, butanol or citric acid. Cassava leaves can be used as a protein vegetable. The fibrous stem can be combusted and used to generate steam and power. The waste water during starch extraction can be subjected to anaerobic fermentation for methane generation, which can be used as a source of power and for heating.

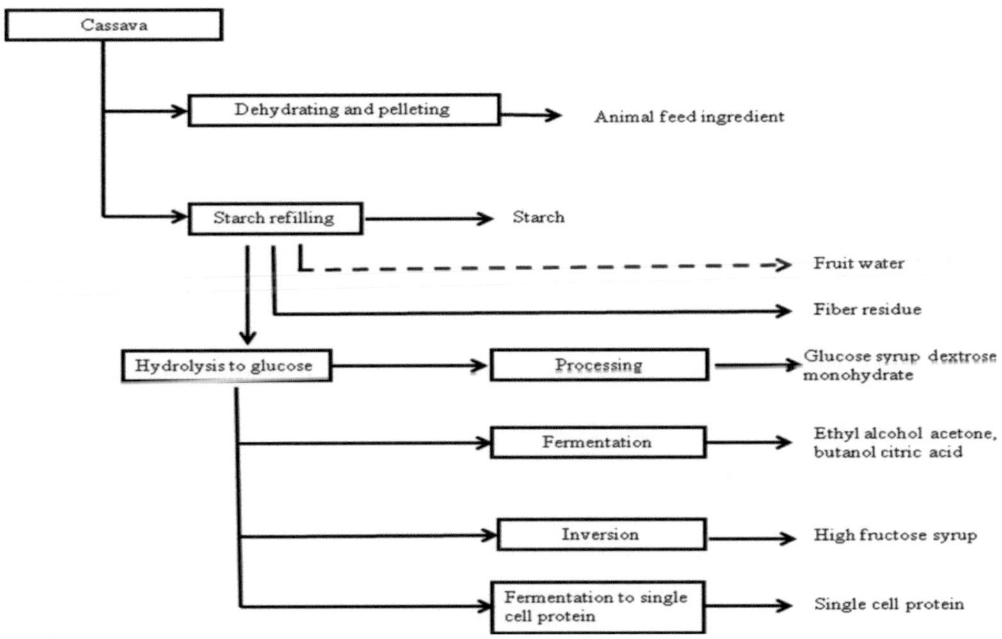

Figure 10. Industrial utilization of cassava.

Sweet potato

Sweet potato (*Ipomoea batatas* (L.) Lam.), a starch-rich root crop, is the third most important (in terms of production) root and tuber crop in Kenya after potato and cassava. It was an important food security crop in the 1960s but its importance has progressively declined with the ascendancy of maize. Nevertheless, sweet potatoes are still valued for their high energy content and high production potential per unit of land. Their cultivation forms part of mixed cropping systems. They are popular food crops in certain regions of the country because the planting material, in the form of the stem, is always available. Sweet potato leaves are used as vegetable and the vines as animal feed. Furthermore, the orange-fleshed variety is a valuable source of beta-carotene, the precursor for vitamin A.

The area planted with sweet potato has steadily increased from about 100,000 hectares in 2000 to 150,000 hectares in 2011 (FAO, 2011). The Lake Victoria basin accounts for about 75% of national production. Sweet potato production ranged between 200,000 and 930,000 tons/year between 1990 and 2011 (Figure 9). The average sweet potato yield in Kenya is about 10 tons/hectare (MoA, 2010b), which is below the genetic potential of 32-37 tons/hectare (FAO, 1990). However, yields are improving due to the introduction of improved cultivars and promotion of better farming practices by Kenya Agricultural Research Institute (KARI) and the Ministry of Agriculture, Livestock and Fisheries. The average national per capita consumption of sweet potatoes is about 25 kg/person/year and is highest around Lake Victoria basin.

Sweet potato roots cannot be stored for more than a few days after harvest because they undergo rapid physiological and microbial deterioration. To overcome this problem, farmers only harvest the mature roots when they are required for consumption or sale and leave the remaining tubers in the ground until when they will be required. The disadvantage is that the mature crop occupies land needlessly which may be used to grow other food crops. The crop is harvested manully by cutting off the vines then digging around the root which is then pulled out of the ground. Initial processing at the farm level involves washing, sorting, grading, storage and bagging. The roots are sold in open-air markets or supermarkets. At home, they are boiled, roasted or fried before they can be consumed. Primary value addition (i.e. washing, peeling, chipping, drying and milling) to obtain flour is rudimentary with limited output. Washing, peeling and chipping are manual on-farm operations. The chips are usually sun-dried after which they are milled, into flours, in *posho* mills (i.e hammer mills), with capacities ranging from 70 to 220 kg/h. The flour is commonly blended with other flours and used to prepare baked goods or porridge. There is a huge unexploited potential for industrial sweet potato processing in the country. They can be processed into chips/crisps for human consumption or used to make animal feeds. Their high starch content makes them valuable ingredients for the starch and starch-by products industries.

Plantain

Plantains (*Musa* spp.) belong to the banana family (Musaceae) and are popularly known as cooking banana because they are starchy, low in sugar and must be cooked prior to eating. Plantain is an insignificant staple food crop in Kenya as compared to neighbouring East African countries. Between 1990 and 2011 production ranged between 15,000 and 35,000 tons in Kenya (Figure 9). During the same period plantain production in Uganda ranged between 8 and 10 million tons/year (FAO, 2011). Production in 2011 was

31,000 tons in Kenya as compared to 10,547,000 tons in Uganda, 729,000 tons in Tanzania and 3,036,000 tons in Rwanda (Figure 11). Per capita consumption of plantains is 19 kg/year in Kenya and 140 kg/year in Uganda (FAO, 2011).

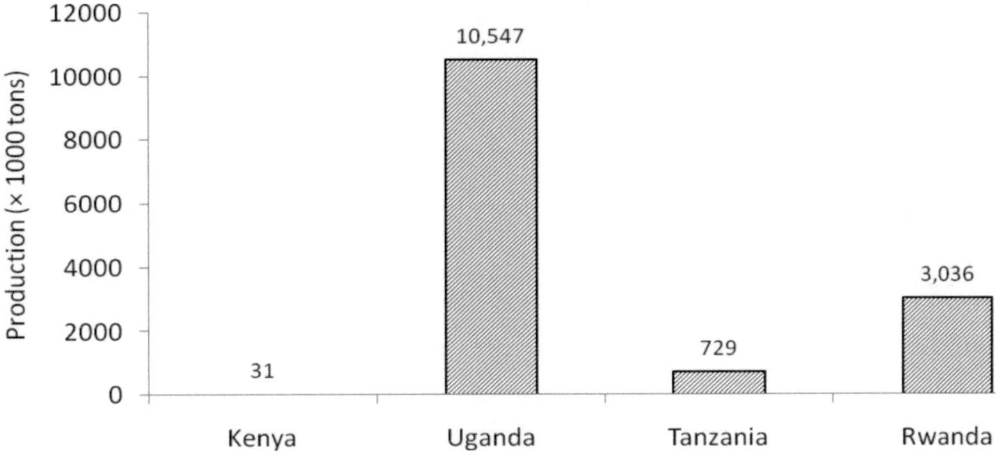

Figure 11. Plantain production in selected countries in East Africa in 2011.

CONCLUSION

Kenya is endowed with diverse agro-climatic zones that can support the production of a wide range of staple fod crops. Generally, national production, processing and consumption of cereals greatly exceed that of roots, tubers and plantains although the latter are still important staple foods in certain regions of the country. Smallholder farmers are the dominant producers of all these crops. Demand for most of these crops is on the rise against declining production and low yields, which are caused by numerous biotic and abiotic constraints. The production deficits of the major staple food crops (i.e. maize, wheat and rice) are met with imports from regional and international markets. Although deficits in the production of the other staple food crops are not covered by imports, it is reasonable to assume that these deficits further contribute to the increased demand for maize, wheat and rice. Post-harvest processing of staple food crops in Kenya is extremely limited (as they are overwhelmingly converted into flour for human consumption) although numerous opportunities exist to convert them into high value food and non-food products.

REFERENCES

ACDI-VOCA – Kenya, Training Manual No. 38. (2010). 2010 Maize Handbook.

Awika JM, Rooney LW. (2004). Sorghum phytochemicals and their potential impact on human health. Phytochemistry, 35:1199-1221.

Bienvenido J. (1999). Comparative nutritive value of various staple foods. Food Reviews International, 15:399-434.

Dixon J, Braun H.-J, Kosina P, Crouch DF. (2009). Wheat Facts and Futures 2009. Mexico, CIMMYT.

FTF (2011). Kenya FY 2011-2015 Multi-year Strategy. Feed the Future. United States Government.

Gitu KW. (2006). Agricultural development and food security in sub-Saharan Africa (SSA). Building a case for more public support. The case of Kenya. A Paper prepared for the Policy Assistance Unit of the FAO Subregional office for East and Southern Africa. Working Paper No. 03.

GoK. (2010). Agricultural Sector Development Strategy 2010-2020. Government of Kenya.

Hammes WP, Brandt MJ, Francis KL, Rosenheim J, Seitter MFH, Vogelmann SA. (2005). Microbial ecology of cereal fermentations. Trends in Food Science and Technology, 16:4-11.

IFPRI (2010). The Aflacontrol Project: Reducing the spread of aflatoxins in Kenya. Project Notes, July 2010.

Jayne TS, Myers RJ, Nyoro J. (2005). Effects of government maize marketing and trade policies on maize market prices in Kenya. Tegemeo Working Paper 15/2005. Tegemeo Institute of Agricultural Policy and Development.
KSC (2012). Kenya Seed Company Ltd. http://www.kenyaseed.com. Accessed on June 27, 2012.
FAO (1990). Roots, tubers, plantains and bananas in human nutrition. Food and Agriculture Organization of the United Nations. Rome.
FAO (1995). Dimensions of need - An atlas of food and agriculture. http://www.fao.org/docrep/U8480E/U8480E00 .htm. Food and Agricultural Organization of the United Nations. Rome. Accessed on 01 April 2013.
FAO (2011). FAOSTAT database. http://faostat.fao.org/. Rome: Food and Agricultural Organization of the United Nations. Rome. Accessed on 01 April 2013.
FAO (2012). FAO Statistical Yearbook 2012. World Food and Agriculture. Food and Agricultural Organization of the United Nations. Rome.
Jones JM, Engleson J. (2009). Whole grains: Benefits and challenges. Annual Review of Food Science and Technology. 1:19-40.
KARI/MIAC (1993). Strategic plan for cereals in Kenya 1993-2013.
Makanda DW, Oehmke JF. (1993). Promise and problem in the development of Kenya's wheat agriculture. USAID Staff Papers.
Mckevith B. (2004). Nutritional aspects of cereals. British Nutrition Foundation Nutrition Bulletin, 29:111-142
MoA (2008). National Rice Development Strategy, Ministry of Agriculture, Livestock and Fisheries, Kenya.
MoA (2010a). Economic Review of Agriculture 2010. Central Planning and Project Monitoring Unit, Ministry of Agriculture, Livestock and Fisheries, Kenya.
MoA (2010b). National Root and Tuber Crops Policy. Ministry of Agriculture, Livestock and Fisheries, Kenya.
Nyoro J. (2006). The compatibility of trade and domestic policy affecting staples in Kenya. Tegemeo Institute of Agricultural Policy and Development.
Odame H, Musyoka P, Kere J. (2008). 'How National Public Policies Encourage or Impede Agribusiness Innovation: Cases of Maize, Tomato and Dairy in Kenya". Country report-Kenya. A World Bank project on "Policies and Incentives for Fostering Technology Development". World Bank, available at http://web.worldbank.org/ WBSITE/EXTERNAL/WBI/WBIPROGRAMS/KFDLP/0.
PREMU (2011). The Drought and Food Crises in the Horn of Africa – Impacts and Proposed Policy Responses for Kenya. Poverty Reduction and Management Unit Africa Region. The World Bank.
USAID (2010). Staple Foods Value Chain Analysis. Country Report – Kenya.
WARDA (2007). Africa Rice Trends. 2007 Brief.
Wortmann CS, Mburu C, Letayo E, Abebe G, Kayuki KC, Chisi M, Mativavarira M, Xerinda S, Ndacyayisenga T. (2006). Atlas of Sorghum. *Sorghum bicolor* (L.) Moench. Production in Eastern and Southern Africa. The Board of Regents of the University of Nebraska on behalf of the University of Nebraska.

2. Genetic modified organisms (GMOs): a brief review of principles, potential socio-economic impact and biosafety challenges in Africa

Linus K'osambo Ayoo[1*]
[1]Kenya Industrial Research and Development Institute (KIRDI), Nairobi, Kenya
[*]*Corresponding author: kosambosan@yahoo.com*

SUMMARY

Application of genetic engineering in the production of genetically modified organisms (GMOs) is one the greatest advances in agriculture since the green revolution. Genetic engineering has resulted in the production of transgenic animals, plants and microorganisms with an array of economically important traits. This technological advancement has promised significant economic impact on livelihood, especially through increasing agricultural productivity. However, introduction of 'foreign' (transgenes) into organisms through genetic engineering has raised fundamental ethical, health and environmental safety questions. This review summarises the main methods of genetic engineering. It outlines techniques such as DNA microinjection, use of (retro) viral vectors, incorporation of genetically manipulated pluripotent stem cells, transfer of genetically altered nuclei into enucleated oocytes, biolistic transformation and *Agrobacterium*-mediated transformation. The main GMO crops and animals that are under production in the world are highlighted. In plant production, the main transgenes that have been engineered into food crops are those that encode herbicide, pests and diseases resistance.

Maize is one of the most important staple crops in Africa. It is also the main relief food crop imported during famine, especially in East, Central and Southern Africa. Specific attention is therefore given to GMO maize, the transgenes engineered and the potential impact of its utilization in Africa. It is known that the main GMO maize cultivars carry the herbicide resistance and/or *cry* insecticidal genes from the bacterium *Bacillus thuringiensis* (Bt). A summary of the current status of scientific support on the safety of GMO maize is outlined. A brief treatise is given on the potential impact of the introduction of Bt maize, biosafety concerns on the effect on the environment especially of non target organisms (NTOs) and the big controversy about GMOs-linked cancer in rats. Pertinent concerns about the development of 'superbugs' as a result of adaptive response of the targeted pests is also briefly discussed. Other scientific controversy issues such as chronic toxicity, divergent biological interpretations and experimental design are mentioned to bring into perspective the core issues of biosafety. The need to carry out long term studies to allay the fears of possible hormonal/sexual/reproductive disorders, carcinogenesis, immune, nervous, genetic and metabolic effects is important for the success of GMOs. A brief theoretical and hypothetical outlook is summarised to address these fears. Finally, the double edged sword of introducing GMOs in Africa is then considered in the context of production systems, technology management infrastructure and environmental concerns. The review concludes by highlighting the key legislations and regulations for the safe utilization of GMOs.

INTRODUCTION

Genetic manipulation through selective breeding and cross-fertilization has been used by man through generations to maintain elite traits in plants and animals. Modern biotechnology especially genetic engineering has revolutionized gene manipulation. The combined use of molecular genetics, DNA recombination and reproductive biology enables the production of transgenic organisms, commonly referred to as genetically modified organisms (GMOs). Production of GMOs involves the introduction of foreign genetic materials (genes) conferring certain traits, such as herbicide resistance or antibiotic resistance into recipient organism. Since the first GM tomato became available in shops in the United States in 1994, dozens of food crops and animals have been modified for greater commercial value, higher yield, improved nutrition or resistance to pests and disease.

There is immense pressure on Africa by GMO advocates who tout bountiful benefits to embrace the gene revolution. African countries such as South Africa, Burkina Faso and Egypt have embraced GMOs (Adenle, 2011) and are currently growing GM crops, while a number, such as Kenya, have enacted regulatory legislations and are poised to start field production and utilization of transgenics. This development urgently requires that the public and policy makers have access to information on the principles and processes of transgenic production, regulation and effect of their utilization on human health and the environment. Pertinent are the issues of biosafety vis-a-vis the anticipated socio-economic benefits of GMOs utilization, especially the capacity of Africa's science and technological infrastructure to manage and significantly gain from genetic engineering.

The introduction of transgenic organisms into the environment, their commercialization and consumption by humans and livestock has kicked off a worldwide debate concerning the safety and regulatory approval process of GMOs (Konig et al., 2004; Seralini et al., 2009; Seralini et al., 2012). Genetic instability associated with cellular natural process of 'rejection' of foreign genes is observed in GMOs (Miller et al., 2002). In addition, the occurrence of recombination 'hotspots' (Kumar and Fladung, 2000; Ho et al., 2000) in transgenic DNA of GMOs brings the risk of horizontal gene transfer for example of antibiotic and herbicide resistance into pathogens and pests. Hepatorenal toxicities as a result of consumption of GMO maize is observed in animal experiments (Vendomois et al., 2009). Concerns exist on the unknown science of the process of gene integration, specifically the risks of insertional mutagenesis, gene silencing and extra-DNA phenomenon in gene expression (IUCN, 2004; IHGC, 2004). The latest controversy involving the apparent cancer-causing GMO maize (Seralini, 2012) raises fundamental questions and a rethink of the health effect of GMOs consumption.

The biosafety concerns as a result of introduction of GMOs into the environment, their consumption and utilization by humans and livestock can be understood by highlighting key theoretical and practical principles of genetic transformation such as *Agrobacterium*-mediated transformation, particle bombardment and gene cloning. Of critical importance are the fundamentals of transgenesis, especially the integration of the transgenes into the genome of GMOs. Of practical concern is the science and protocols underlying the animal studies and experiments on the effects of GMOs consumption. Equally relevant is the prospect of development of superbugs as a result of adaptive response and evolution of the targeted pests and pathogens (Tabashnik, 1994). The socio-economic benefits within the context of African agricultural production systems need to be weighed. The pitfalls of monoculture as a result of dependence of a few high performing GMOs especially on genetic diversity and insurance against vicissitude of nature should be addressed.

This brief review gives a summary of the potential benefits of GMOs in Africa. To understand the biosafety concerns, it outlines and discusses the basic concepts of GMOs production. The review briefly discusses the main biosafety issues relevant to health and environmental effects such as GMOs gene stability, extra-DNA phenomenon, insertional mutagenesis, horizontal gene flow and emergence of super bugs. Socio-economic impact of GMOs and gene revolution in Africa is briefly considered in the context of the continent's complex production system and global food policy. The review concludes by summarising Kenya's and international legislations enacted to safeguard biodiversity and biosafety.

Principles of GMOs production

Genetic engineering is the umbrella term for procedures that result in a directed alteration in the genotype of an organism. The combined use of molecular genetics, DNA recombination and reproductive biology enables the generation of transgenic organisms, commonly referred to as genetically modified organisms (GMOs). Production of genetically modified animals, plants and microorganism are fundamentally similar and generally involves two main routes: the direct introduction of foreign genetic material and use of vectors such as viruses and bacteria to transfer the genetic material in specially constructed plasmids. The main route in animals include: (i) microinjection of DNA into the pronucleus of a fertilized oocyte (zygote) (Hammer et al., 1985; Brem et al., 1985; Wall, 2002; Webster et al., 2005), (ii) integration of a (retro)viral vector into an oocyte or an early embryo (Chan et al., 1998), (iii) incorporation of genetically manipulated pluripotent stem cells into an early embryo (Capecchi, 2000), and; (iv) transfer of genetically altered nuclei into enucleated oocytes (cloning) (Willadsen, 1986; Wells et al., 1999; Betthauser, 2000; Woods, 2003). In plants, the predominant methods are (i) *Agrobacterium* transformation (Tzfira and Citovsky, 2002; Gelvin, 2000) (ii) transformation of protoplasts (Potrykus, 1995) and (iii) particle bombardment (Cheng et al., 1997; Zhao et al., 2000; Kosambo-Ayoo et al., 2011). This review will focus mainly on the production of genetically modified plants.

In vector-mediated transformation via *Agrobacterium*, the natural ability of this bacterial phytopathogen to transfer DNA into plant cells is exploited. The molecular basis of genetic transformation of plant cells by *Agrobacterium* is the transfer and integration of a region of a large tumour-inducing (Ti) or rhizogenic (Ri) plasmid resident in this bacterium into the plant nuclear genome (Stam et al., 1997; Gelvin, 2003). During infection, a specific segment of the plasmid DNA, referred to as T-DNA, is inserted into the plant genome (Stirn and Lorz, 2006). The T-DNA contains genes of phytohormones which are responsible for cell proliferation and for formation of the crown gall, and genes for the formation of special nutrients (opines) in the plant cell (Hansen, 1999). When *Agrobacterium* is used as a tool in genetic engineering these genes are deleted and replaced by the genes of interest. Advantages of this method include the typical insertion of only one or a few copies of the transgene (Dai et al., 2001) and the transfer of relatively large segments of DNA with only minimal rearrangements (Arencibia et al., 1998).

Direct gene - includes DNA uptake into protoplasts (protoplast transformation) and shooting of DNA-coated particles into tissues (particle bombardment). Plant cell wall is a barrier to gene transfer. One way of circumventing this barrier is to use plant cells of which the cell wall has been digested enzymatically, resulting in so-called protoplasts. DNA uptake into protoplasts can be stimulated by the use of either poly(ethylene glycol) (PEG - transformation) or electric pulses (electroporation). When an appropriate protoplast-to-plant regeneration system is available, large numbers of transformed clones can be regenerated to afford fertile transgenic plants. Particle bombardment involves the use of high velocity microprojectiles to deliver DNA into plant tissue. A particle gun is used to accelerate DNA-coated microprojectiles into cells, past the cell wall and the cell membrane. Particle bombardment and other direct DNA uptake methods often results in complex insertion loci, which can cause gene silencing (Hammond, 1999). In respect to biosafety, *Agrobacterium*-mediated transformation is therefore relatively safer than the projectile method.

Genes used in the development of GM plants

The main biosafety risks emanates from the introduction of foreign genes that include the primary gene(s) of economic interest (such as disease resistance gene) and the accompanying genes in the plasmid (or gene construct) used during gene cloning and tissues transformation. These additions are dictated by the molecular requirements for gene expression. A typical plant gene consists of a promoter, a coding sequence, a transcription terminator, and a polyadenylation signal. The promoter is the main determinant of the expression pattern in the transgenic plant. The promoter directing the synthesis of the cauliflower mosaic virus 35S RNA is the most frequently used constitutive promoter. For monocots, promoters from the rice actin 1 gene (*act*1) or the maize ubiquitin 1 gene (*ubi*1) are used (Fütterer, 1995; Christensen et al., 1992). Other promoters used such as *CaMV* have raised biosafety issues as it represents a recombination hotspot in transgenics (Windels, 2001; Ho, 1999; Ho, 2000).

Selection marker genes are normally included in the gene construct used in transformation to enable selection of transformed cells due to the traditionally low efficacy of stable gene transfer. The most commonly used selectable marker genes in transgenic plants code for antibiotic or herbicide resistance. The main ones are: (i) *neo* or *nptII* gene isolated from transposon Tn5 from *Escherichia coli* K12 codes for neomycin phosphotransferase (NPTII) that confers neomycin, kanamycin, and geneticin resistance (Collard, 1999), (ii) *hpt* gene isolated from *E. Coli* codes hygromycin phosphotransferase that confers hygromicin B resistance (Chrott 1995), (iii) *bla* gene codes for TEM-1 beta lactamase for ampicillin resistance (Moens et al., 2002), (iv) *aadA* gene confers resistance to streptomycin and spectinomycin (Schrott, 1995). The main herbicide resistance genes are (i) *bar/pat* gene: the bar gene from *Streptomyces hygroscopicus* and the *pat* gene from *S. Viridichromogenes* code for phosphinotricin acetyltransferase (PAT) which confers resistance to the herbicide compound phosphinotricin (Block et al., 1987), (ii) EPSPS gene codes for the enzyme 5-enolpyruvylshikimate-5-phosphate synthase (EPSPS) confers resistance to non-selective herbicide glyphosate (active ingredient of the commercially available herbicide Roundup) (Stirn and Lörz, 2006). Reporter genes are also used to select transformed cells. Reporter genes code for products which can be detected directly or catalysed reactions whose products are detectable. The two main reporter genes are the *gusA* and *luc* genes whose products can be detected through spectrometry and bioluminescence respectively (Jetterson et al., 1987; Ow et al., 1986; Miller, 1992). The selection and marker genes may end up in the GMOs produced and present the greatest danger of horizontal gene transfer as most of them are microorganism-based.

GM plants in the market

By 2006, there were 23 GM (crop) plants or products thereof that had been approved for commercial use in Europe (Stirn and Lorz, 2006) (Table 1). The number is higher today. Approximately 72% of GM plants contain herbicide resistance gene, followed by 19% carrying insect resistance genes and 9% carrying both herbicide and insect resistance genes (stacked genes). The main categories of GM plants on the market are: (i) herbicide resistance in soybean, maize, oilseed rape, sugar beet, wheat, rice and cotton, (ii) insect resistance in maize, potatoes, tomatoes and cotton, (iii) virus resistance, male sterility, delayed fruit ripening and fatty acid content.

Table 1. A selection of GMOs approved in the EU showing the traits (genes), the applicants and approved use.

	Plant (line)	Trait (gene)	Applicant	Approved use	Year
1	Soybean (GTS-40-3-2)	HR (CP4 EPSPS)	Monsanto	Import and processing	1996
2	Maize (Bt 176)	IR (cryIAb), HR (bar), ABR (bla)	Ciba-Geigi	Cultivation; import and processing	1997
3	Maize (Bt 11)	IR (cryIAb), HR (pat)	Northrup King	Import and processing; food use	1998 2004
4	Maize (MON 810)	IR (cryIAb)	Monsanto	Cultivation; flour, semolina, starch and starch products, oil	1998 1997
5	Maize (T15)	HR (*pat*), ABR (partial *bla*)	AgrEvo	Cultivation, refined oil, starch and starch products, fermented or heat-treated products from maize flour	1998
6	Maize (NK 603)	HR (CP4 EPSPS)	Monsanto	Import and processing	2004
7	Maize (MON 863)	IT (cryIIIBb2), ABR (nptII)	Monsanto	Import and processing	2005
8	Maize (GA 21)	HR (maize EPSPS)	Monsanto	Processed food	2005
9	Maize (TC 1507)	IR (cryIFa2), HR (pat)	Mycogen/Pioneer	Food and feed use	2005
10	Oil-seed rape (MS1xRF1; MS1xRF2)	MS/RF (barmase/barstar), HR (bar), ABR (nptII)	PGS	Cultivation (breeding purposes); refined oil	1997
11	Oilseed rape (MS8cRF3)	MS/RF (barmase/barstar), HR (bar)	PGS	Refined oil	1999
12	Oil-seed rape (HCN 92 = Topas 19/2)	HR (pat), ABR (nptII)	AgrEvo	Import and processing (feed); refined oil	1997
13	Oil-seed rape (HCN 28 = T45)	HR (pat)	AgrEvo	Import and processing	2005
14	Oil-seed rape (GT 73)	HR (CP4 EPSPS, gox)	Monsanto	Refined oil	1997
15	Cotton (MON 1445/1698)	HR (CP4 EPSPS), ABR (nptII, aad)	Monsanto	Refined oil	2002
16	Cotton (MON 531/757/1076)	IR (cryIAc), ABR (nptII, aad)	Monsanto	Refined oil	2002
17	Cotton (MON 15985)	IR (cryIAc, cryIAb), ABR (nptII, aad), RP (uidA)	Monsanto	Refined oil	2002

Benefits of GMOs

It is anticipated by the proponents of GMOs that genetic engineering will overcome the productivity constraints of conventional plant breeding, that novel plant and animals with desirable agronomic and nutritional qualities will be produced to address global deficiencies (Giannesi and Carpenter, 2000) and that transgenic crops will reduce pesticide use (Zilberman et al., 2007). It is projected that GMOs will boost food security for the world's growing population by raising sustainable food production. Use of GMOs will benefit the environment by reducing the need for more farmland, irrigation and pesticides. It will also provide better medical treatments and vaccines, new industrial products and improved fibres and fuels. GMOs are expected to increase agricultural/maricultural productivity, maximising per hectare and per capita yields. GMO use also offers the potential for development of "issue-specific solutions" to problems facing particular communities, such as the advent of a new pest or disease. Carbon-storage and climate change benefits may accrue from the use of GM trees.

Directed economic benefits accruing from GMOs utilization in Africa has been reported in countries that have embraced gene technology. It is reported that of the 15.4 million farmers that planted GM crops in 2010, over 90% (14.4 million) are resource-poor farmers in developing countries, including three African countries (Burkina Faso, South Africa and Egypt) (James, 2010). For example, almost 100,000 farmers in Burkina Faso cultivated GM cotton on 260,000 hectares in 2010, which represents 126% increase from 115,000 hectares planted in 2009 (Adenle, 2011). In terms of contribution to the economy of Burkina Faso, it is estimated that a benefit of over US$ 100 million per year could be generated based on almost 30% increase in yield (Vitale, 2008). Additionally, close to a 50% reduction in insecticides sprays may be realised (Vitale, 2008). The economy of South Africa benefited a great deal from the adoption of GM technology and it is reported to have enhanced farm income from GM maize, GM soybean and GM cotton by US$ 156 million in the period 1998 to 2006, with 2006 alone estimated at US$ 67 million (James, 2007).

GMOs biosafety

Genetic engineering and the production of GMOs raise tangible environmental, health and socio-economic issues. Because genetic engineering is relatively new, much is still unknown about how its products may behave and evolve, and how they may interact with other species. Pertinent questions are raised about the utilization of GMOs: Could an ability to tolerate herbicides, for example, transfer from GM crops to related wild species? Might plants that have been genetically modified to repel pests also harm beneficial insects? Could the increased competitiveness of a GMO cause it to damage biologically-rich ecosystems? Risks of environmental contamination have been actualised, for example, in instances where modified soya has been found in export shipments that had been declared GMO free, or pollen from modified corn is detected in a neighbouring non-modified field. Additionally, a number of biosafety issues such as the implication of unknown science due to extra-DNA phenomenon, insertional mutagenesis, food allergens, toxins, hepatorenal toxicities of herbicide residues, metabolic interferences and horizontal gene flow are pertinent in addressing the biosafety of GMOs. In the back drop of the raging debate on the biosafety of GMOs, several international regulatory bodies including the World Health Organisation (WHO), the Food and Agriculture Organisation of United Nations (FAO) have concluded that there is no scientific evidence that the application of GM technology has resulted in substantial human health effects or environmental problems (Paarlberg, 2010; RS, 2003). Nonetheless, it is imperative to systematically review the contentious biosafety issues.

GMO maize

Maize (*Zea mays*) is an important staple in Africa. It is also one of the main relief food shipped to vulnerable populations during famine. Kenya has given approval to huge shipment of GM maize as relief food. There is every probability that these shipments can find their way into the local market and into farmers' hands before approval for local consumption is procedurally approved. In this scenario, urgent public awareness of the biosafety and risks of GM maize is critical. It is important to avail information on the biosafety of GM maize, specifically, the genes used, the effect of novel transgenic metabolites and herbicides residues on human health and effect on non-target organisms.

The GM maize in the market are transformed with herbicide resistance genes and/or *Bacillus thuringiensis* (*Bt*) genes encoding pest resistance. Most of the *Bt* maize varieties express either *Cry1Ab* gene for the control of stem borers, such as the European corn borer, *Ostrinia nubilalis* (Hubner) (Lepidoptera:Crambidae) (events MON810 and Bt11), or *Cry3Bb1* for the control of corn rootworms (*Diabrotica spp*.; Coleoptera: Chrysomelidae) (events MON863 and MON88017) (Hellmich et al., 2008; CERA, 2010). Some GM maize like NK 603 has been genetically engineered to tolerate the broad spectrum herbicide Roundup and thus contains residues of this chemical. There is heated international debate touching on the biosafety of GM maize, especially its effect on human health and non-target organisms and socio-cultural impact of its utilization. Of more critical concern are the findings that certain GM maize have hepatorenal toxicities, in addition to other suspected side effects on human health. Analysis of regulatory raw data from GM maize NK 603, MON 810 and MON 863 show that hepatorenal toxicities are possible (Vendomois et al., 2009). A toxic effect is mostly concentrated in kidney and liver function, the two major diet detoxification organs (Vendomois et al., 2009). In addition, some effects on heart, adrenal, spleen and blood cells are also noted (Vendomois et al., 2009). The toxicity is attributed to the new pesticides (herbicide or insecticide) present specifically in each type of GM maize. However, these results are dismissed as biologically insignificant (Doull et al., 2007). The toxic effects reported, though isolated, should be treated with due importance. Further studies should be undertaken to disapprove or clarify the implied risks to human health.

The effect of GM maize on a non target organism (NTO) is illustrated in a report on the potential negative of the *Cry1Ab* protein on larvae of the two spotted ladybird *Adalia bipunctata* (Schmidt et al., 2009). However, this report has been disputed to the effect that the observed harmful effects were artefacts of poor study design and procedures (Alfageme et al., 2010; Rauschen, 2010).

The recent global panic and outcry after the report on the cancer-causing GM maize (Seralini et al., 2012) is illustrative of the knowledge chasm that needs to be filled before all fears are allayed on the biosafety of GMOs. Seralini et al. (2012) studied the health effects of a Roundup-tolerant genetically modified maize (NK603) in rats for 2 years. Sex hormone modification, liver congestion and necrosis, severe kidney necropathies, large mammary tumours in female and male rats and deaths associated with severe hepatorenal insufficiency were observed as a result of consumption of GM maize (Seralini et al., 2012). The European Food Safety Authority (EFSA) however maintains that the conclusion of Seralini et al. (2012) raising concerns about the potential toxicity of GM NK603 and of a herbicide containing glyphosate is of insufficient scientific quality to be considered as valid for risk assessment. Central to this is the conclusion that the strain of rat (Harlan Sprague Dawley strain) used in the Seralini et al. (2012) study is prone to developing tumours during their life expectancy of approximately two years, therefore the observed frequency of tumours is influenced by the natural incidence of tumours typical of this strain, regardless of any treatment

The above biosafety concerns, although dispelled in subsequent analyses and studies, demonstrate the need to carry out exhaustive long term risks assessment and embrace the principle of open information and data access, especially toxicological information held by the transnational organizations and regulatory bodies involved in GMOs development and release.

Genetic stability

Genetic stability measured by the inherent capacity of a trait or transgene in a plant to be inherited through generations is important in predicting the risks and benefits of GMOs. However, there are mechanisms in the cell that actively seek out, inactivate or eliminate foreign DNA from the genome (Miller et al., 2002). There is new evidence that the transformation process for making transgenic plants induces further genetic instability (Kumpatla, 1998; Horvath, 2001; Svitashev, 2000) leading to chromosomal rearrangements and genome scrambling. Moreover, genetic stability as a uniform line is the single most important criterion for approval of a GMO for commercialization as exemplified by the European Commission (EC) test for Distinctiveness, Uniformity and Stability (the DUS test). The EC Directive on deliberate release requires strict molecular evidence of genetic stability, which is also necessary for establishing the identity of the transgenic line and to ensure traceability. It is difficult to determine environmental or health risk unless a crop can be identified and maintained as a genetically stable entity. From the biosafety point of view, this structural instability, the tendency for the transgenic DNA to come loose, to rearrange or become lost in part or in whole in successive generations (Ho, 1998; Ho, 1999; ISIS, 1999-2001) is critical. It can therefore be concluded that this genetic instability could change the transgenic line in unpredictable ways in terms of health and environmental risks.

Extra-DNA phenomenon beyond Watson-Crick

The scientific foundation of GMOs production and by implication its risk assessment is the basic principle that DNA is the building block of life. However, Human Genome Project (IHGC, 2004) came to the conclusion that DNA is not sufficiently varied and does not allow a sufficient number of combinations to account for all biologically replicated traits, even of simpler life forms. This suggests that there are other factors which are also "building blocks" of life. Statistical information concerning the number of GMOs which fail to show the expected characteristics, or which show new characteristics and other types of instability is not supported by the theory of DNA as the basic blueprint of life. These observations pointing to extra-DNA phenomenon leads to an important conclusion: If Watson-Crick postulation of the process of GMO creation is incorrect, then it is impossible to evaluate the risks of GMOs, without first resolving the underlying principles.

Insertional mutagenesis, allergies and immune depression

During genetic engineering and transformation, plasmid constructs containing desired DNA are introduced into the genome of the targeted ex-plant. The process of insertion into the genome is mostly random, more so if biolistic transformation method is employed. Insertion of a novel gene can have an unintended impact on the rest of the host's genome that results in unforeseen side effects. Researchers note that GM varieties exhibit traits not expected by virtue of the specific gene replaced. In 2000, Monsanto's Roundup Ready soya, commercially grown for years, was analysed by molecular methods. Not only is the gene order of the insert found to be scrambled, the plant genome at the site of insertion is also scrambled, and there is a 534 base pairs fragment of unknown origin in there as well (Tax and Vernon, 2001). Mustard seeds engineered for herbicide resistance were also found to be twenty times more fertile than their non-GM equivalent. This is an illustration of insertional mutagenesis.

Novel transgenic proteins produced by GMOs can be allergic. A GM soybean that contained a Brazil nut allergen was reported by Noodle (1996). The intense media coverage resulted in cessation of further development of this particular variety due to potential health risks. Another incident that attracted public attention was that of a scientist that claimed a genetically engineered potato caused a depressed immune response in rats (Owen and Pusztai, 1999) but subsequent investigation showed that the experiment was scientifically flawed.

Horizontal gene flow

Horizontal gene flow refers to exchange of genetic material between species. There are contrasting reports of horizontal gene flow between GMOs and their wild relatives. Experiments have show that the rate of cross-pollination between conventional and GM varieties of potatoes are generally low and become negligible when the separation distance exceeds 10 metres (Rogers, 1995). By contrast, Danish field trials have shown that oilseed rape modified for herbicide tolerance can easily cross with wild brassica species such as wild mustard (Chevre, 1997). Consequently, cross-pollination between GM and non-GM oil seed rape has been detected at distances of up to 2 km. Recently, landraces of corn growing in remote regions of Mexico were found contaminated with transgenic corn DNA by probing with the *CaMV 35S* promoter (Quist and Chapela, 2001). Molecular analysis showed that the sequences next to the promoter are very diverse, as consistent with horizontal gene transfer and recombination (Ho, 2001).

Transgenic DNA typically contains DNA from widely different sources, mainly bacteria and viruses and other genetic parasites that cause diseases and spread antibiotic resistance. There exists the potential for homologous recombination with those agents due to similarities in DNA base-sequences. Homology enhances horizontal gene transfer 10 million to 100 million-fold (Old and Primrose, 1994). In addition, certain 'receptive hotspots' exist in both the plant (DeVries et al., 2001) and the human genome (Kumar and Fladung, 2000). These may also be 'recombination hotspots', prone to breaking and rejoining. That would mean inserted transgenes are more likely to be lost, to recombine, or to invade other genomes. Cauliflower mosaic virus (*CaMV*) *35S* promoter is regarded as a recombination hotspot (Ho et al., 2000) that could enhance horizontal gene transfer and recombination (Windels, 2001; Ho, 1999; Ho, 2000). *CaMV 35S* promoter is active in species across the entire living world, including frog eggs and human cells (Ho et al., 2000). *CaMV 35S* promoter, if transferred to human or animal cells, could activate cancer-associated genes as well as dormant viruses that are in all genomes.

Super bugs and susceptibility of GMOs to pests and diseases

It is observed that *Bt* cotton are more susceptible to the fungal disease *Fusarium* wilt than conventional cotton. Forty years of empirical evidence from the USA, Japan, Central America and China demonstrates that thc usc of thc pcsticidcs consisting of Bt toxin has allowed some agricultural pests (such as the diamond back moth *Plitella xylyostella*) to evolve distinct toxin resistant populations (Tabashnik, 1994).

Experimental design and data on chronic effects of GMOs and products

Nutritional tests with weight, bone mass, milk or meat production are available. Acute toxicological tests with recombinant proteins, *in vitro* digestibility of transgenic proteins and limited compositional analysis among other data are also available. However, possible chronic side effects of pesticide residues are not scientifically assessed. Most regulatory GMOs tests are 3 months long feeding trials of laboratory rats, biochemically assessed. Tests are not compulsory, are not independently conducted and test data mainly confidential. The design, analyses, interpretation and application of animals feeding studies should be comprehensive to allay the fears of subjective interpretations. In order to address the issue of biosafety, it is necessary to have access to toxicological tests, preferably on mammals performed over longest time-scale involving detailed blood and organ system analysis performed in accordance with international standards.

Labelling, traceability and epidemiological studies

Labelling of GMOs provides clear and factual information to the consumer for informed choices. In absence of labelling and traceability, there can never be any human or animal epidemiological studies to support safety claims. Lack of labelling and traceability in GMO-producing countries has made epidemiological studies impossible. Over 97% of edible GMOs among cultivated GMOs (soy, corn, oilseed rape) are grown in South and North America, where GMOs are not labelled. Kenyan millers are againsti the regulations on labelling of GM foods. However, within the European Union, products that are GMOs or derived thereof, in whole or in part should be labelled as such (European regulation 1829/2003). Labelling makes good sense in understanding the acute and chronic side-effects, especially after generations of consumption of a new and an unnatural product.

GMOS AND AFRICA

Threat to biodiversity and epidemics

Africa is the origin and home of rich gene pool of flora and fauna. Biodiversity in Africa is under immense pressure because of impoverization and overdependence on natural resources for survival. Globally, it is estimated that 40% of the world economy is derived from biological diversity. Humanity is pushing ecosystems, species and gene pools to extinction faster than at any time since the dinosaurs died out 65 million years ago. More than 31,000 plant and animal species are threatened with extinction and at least one breed of livestock dies out every week. It is posited that GMOs and gene technology is a threat to biodiversity by fronting monoculture, control of seed systems by a few corporations, discouraging traditional collection and preservation of seeds over seasons and socio-economic pressure on biodiversity friendly organic farmers. Widespread planting of genetically uniform GM crops under monoculture conditions can rapidly displace local varieties with greater genetic potential to resist diseases. This can set the stage for epidemics.

Socio-cultural impact

The experience of other countries shows that that leaping towards genetic engineering brings challenging socio-economic impacts. It requires the acceptance of intellectual property rights on living organisms, privatisation of public research and expensive research and development on biotechnology and auxilliary support systems. The above diverts badly needed funds from the government coffers to the detriment of farmer-based innovations. It is apparent that evidence linking benefits to GM use is limited, and often provided only in episodic form. Agricultural statistics are difficult to find that provide appropriate linkages between GM crops and productivity. Claims that varieties can be developed more quickly with GM techniques than through more traditional methods are also not entirely supported by available facts. It is therefore safe to conclude that the taunted agricultural benefits are either exaggerated or not well documented.

Concerns that over time non-GM varieties, which along with their wild relatives are the basis on which GMO development is founded will begin to disappear. This may happen through voluntary action, where farmers may abandon traditional varieties. It may also occur involuntarily, where pesticide-ready or pest-resistant crops affect neighbouring non-GMO fields by altering pest patterns (such as increasing stress on non-GMO crops) or affect the established system that includes the pest species (such as birds and other creatures that feed on insect populations or larvae). It may also result from genetic contamination, where hitherto organic food market may reject produce from regions with GMOs.

Sustainable agriculture in complex production systems: Green Revolution vs Gene Revolution

It is widely held that green revolution technologies did not bypass Africa, rather they were available but unpopular and ineffective. The lesson learnt is that " breakthrough" technologies, can only have limited success in Africa's complex ecology. African soils are generally unsuitable to intensive, monoculture production because of insufficient or excessive rains, high incidence of diseases and pests, and other factors. Proper agricultural management requires a much more complex approach. Farmers in Africa lack access to markets, infrastructure, research extension services, and all other forms of support. Women, who constitute the majority of farmers in Africa, are often left to manage their households and their farms with few resources, as the men look for wage labour far from home. Costs of inputs, especially purchase of transnational organizations controlled seeds and herbicides, for GM crops is likely to make production expensive. Lessons can also be drawn from the assimilation of the green revolution in Africa. In Kenya, the National Farmers Union says that market liberalisation drove up prices of inputs, forcing many small farmers to move back from Green Revolution to subsistence agriculture. It is noted that technology is a relatively insignificant constraint in African agriculture. If farmers had the incentives and conditions to invest in farming, Africa could easily take care of its food security. Natural resources available in sub-Saharan Africa are "grossly under-utilised", as the continent only produces 0.8% of its potential agricultural yields. It is argued that the "main obstacles to increased crop production are socio-economic." GMOs social and economic concerns and new biosafety risks require considerable resources to manage. It shifts control over agricultural research and development towards transnational organisations and constrains the collective nature of plant breeding that has existed since time immemorial. Gene Revolution gives a second wind to the failed Green Revolution model paradigm.

In contrast to the green and gene revolution paradigm, sustainable agriculture offers an entirely different approach to agricultural development. It encourages development within agricultural systems to eliminate, non-renewable external inputs such a chemical fertilisers and pesticides. The technologies and practices mimic natural ecosystems, such as multiple cropping and valley cropping systems. It encourages principle of equity whereby farmers have access to seeds, knowledge and other resources. Indigenous knowledge systems and biodiversity are the foundations of sustainable seed systems. Farmers are active participants in plant breeding, from genetic conservation and crop improvement to the marketing and distribution of seeds. Scientists work alongside farmers to strengthen and support their breeding strategies. While genetic engineers aim to find solutions exclusively at the gene level, practitioners of sustainable agriculture look at soil health, water management and crop combinations. Sustainable agriculture embodies complexity and diversity, while genetic engineering is based on simplicity and uniformity.

Global agricultural policies – subsidies and food betting

Agriculture is the backbone of the economies of most sub-Saharan African countries. Agricultural exports from these countries are not competitive in the international market mainly because of subsidies. The EU's Common Agriculture Policy (CAP) has far reaching effect on global food market. While only 5% of Europeans work in agriculture, around half of the European Union's budget is spent on subsidizing farmers through CAP resulting in a massive waste of resources, high food prices and an adverse impact on agricultural producers in developing countries. Subsidies push down farm produce prices and cause international dumping. In 2010, the EU spent €57 billion on agricultural development, of which €39 billion was spent on direct subsidies. The United States currently pays around US$ 20 billion per year to farmers in direct subsidies as "farm income stabilization. It is the extraordinary distortion of global trade where the West spends US$ 360 billion a year on protecting its agriculture with a network of subsidies and tariffs that costs developing countries about US$ 50 billion in potential lost agricultural exports. The policy of agricultural subsidies have far much more damaging effect on African agricultural economies, far much more than can be redeemed by the socio-economic benefits that can possibly accrue from wholesale acceptance and utilization of GMOs.

Food betting is another global issue that negates the benefits that can be gotten from the utilization of GMOs. Deregulation in the late 1990s allowed financial institutions to bet on food prices, resulting in some US$ 200 billion being poured into the market. For example, hedge fund Armajaro virtually single-handedly sent the global price of cocoa to a 33-year high in July 2010 by buying around 15% of global cocoa stocks. Others seem to be yielding to pressure. Last year (2012), several German banks, including the second largest, Commerzbank, ceased to speculate on basic food prices for moral reasons.

Another relevant global phenomenon and policy trend is where the donor community pegs provision of assistance (direct food aid and famine relief) on the condition that the recipient country is willing to accept GMOs. This has laid bare overt manipulation of 'starving nations' into accepting GMOs. Critics have lamented that vested interests intentionally use a time of national crisis as a tool for forcing governments to take action against national principles and public opinion.

GMOs and Intellectual Property Rights

The driving force behind the commercialization and utilization of GM crops are the transnational organizations who have heavily invested in GMOs development as well in the seed, pesticide and herbicide industry. GMO development is expensive and transnational organisations are poised to protect their intellectual property rights and recover their investments in GMO research and development. Legal battles touching in intellectual property rights have pitted transnational organisations against farmers who have in indavertently replanted GMOs seeds from preceding crops. Seed giant Monsanto won more than US$ 23 million from hundreds of small farmers accused of replanting the company's genetically engineered seeds

(RT, 2013). The lawsuits concern Monsanto's patent rights as the company strives to prevent farmers from replanting crops grown from the company's seeds. It is a concept that a recent study – titled 'Seed Giants vs. US Farmers' referred to as creating a "seed oligarchy." In the report, the Center for Food Safety (CFS) said it discovered 142 patent infringement suits against 410 farmers and 56 small businesses in more than 27 States as of December 2012.

LEGISLATIONS

Cartagena Protocol

In 1995, the Parties to the Convention on Biological Diversity responded to biosafety challenge by launching negotiations on a legally binding agreement that would address potential risks posed by GMOs. These discussions culminated in January 2000 with the adoption of the Cartagena Protocol on Biosafety. It is named after the Colombian city where the final round of talks was launched. The Protocol sets out a comprehensive regulatory system for ensuring the safe transfer, handling and use of GMOs subject to trans-boundary movement. In this way, the Protocol seeks to meet the needs of consumers, industry and the environment. To promote biosafety, the Protocol reflects a fundamental concept known as the precautionary approach. This concept means that a government may decide on the basis of precaution not to permit a particular GMO to be imported across its borders. The Protocol applies precaution not just to biodiversity, but to potential risks to human health as well. It also gives importing countries the right to take into account socioeconomic concerns (provided their actions are "consistent with their international obligations"). Such concerns could include the risk that imports of genetically engineered foods may replace traditional crops, undermine local cultures and traditions or reduce the value of biodiversity to indigenous communities.

National biosafety regulations

A number of African countries, such Kenya, Egypt, Burkina Faso, South Africa and Nigeria have enacted Biosafety Acts and are poised to utilize GMOs. The Biosafety Act (2009) of Kenya received Presidential assent on 12th February, 2009. The Act, developed within the spirit of the Cartagena Protocol, lays down the legal and institutional framework to govern biotechnology in Kenya. Prior to enactment of the Biosafety Act (2009). Kenya had only biosafety regulations and guidelines lacking backing of an Act of Parliament, even though Kenya signed the Cartagena Protocol for Biosafety in 2000 and ratified it in 2003. Kenya also has now in place a National Policy on Biotechnology and Biosafety. Kenya has enacted a set of legislation for the utilization of GMOs. Central to this is the National Biosafety Committee whose membership includes, National Council for Science and Technology (NCST), Kenya Plant Health Inspectorate Service (KEPHIS), International Livestock Research Institute (ILRI), Ministry of Agriculture, Livestock and Fisheries, University of Nairobi, Kenya National Federation of Agricultural Producers (KENFAP), Africa Biotechnology Stakeholders Forum (ABSF), Kenya Industrial Property Institute (KIPI), Kenya Medical Research Institute (KEMRI), Kenya Plant Breeders Association, Ministry of Public Health and Sanitation, Consumers Information Network, Seed Trade Association of Kenya, Ministry of Higher Education, Science and Technology, Kenya Bureau of Standards (KEBS), National Environment Management Authority (NEMA), Department of Veterinary Services and Kenya Agricultural Research Institute (KARI).

CONCLUSION

Genetic engineering has the potential of solving the problem of food insecurity. However, concerns of biosafety have limited the assimilation and application of this novel technological development. The assurances of world food safety organizations such as FAO/WHO and European Food Safety Authority that GMOs consumption is safe has met concerted rebuttal from a section of scientific community and NGOs who enumerate health, environment and negative socio-economic impact of their introduction. Sober public dialogue replete with well documented scientific and verifiable evidence is important in allaying the fears of the public and policy makers. The relevance of GMOs in complex African agricultural production system should be critically analyzed lest it be an expensive perpetuation of the ineffectual silver bullet paradigm of the green revolution. Morbid fear of the transnational organizations especially their control of GMOs production, seed, herbicide and pesticide industry should be dispelled by transparency in the development and GMO approval processes. Development and application of strong national regulations and concomitant institutional and technological support system is important in safely harnessing genetic engineering.

REFERENCES

Block M, Botterman J, Vandeweile M, Dockx J, Thoen C, Gossele V, Movva NR, Thompson C, Van Montague M, Leemans J (1987). Engineering herbicide resistance plants by expression of a detoxifying enzymes. EMBO J. 6:2513-2518.

Collard JM (1999). Aminoglocoside resistance. http://www.antibioresistance.be/aminoglycosides.html (accessible February 2013).

Devlin K (2002). Genetically modified crops in Africa: Implications for small scale farmers. Genetic Resource Action International (GRAIN) 1-24.

DeVries J, Meier P, Wackernagel W. (2001). The natural transformation of the soil bacteria. *Pseudomonas stutzeri* and *Acinetobacter* sp. by transgenic plant DNA strictly depends on homologous sequences in the recipient cells. *FEMS* Microbiology Letters, 195:211-5.

Doull J, Gaylor D, Greim HA. (2007) Report of an Expert Panel on the reanalysis by Séralini et al. (2007) of a 90-day study conducted by Monsanto in support of the safety of a genetically modified corn variety (MON 863). Food Chem Toxicol. 45:2073-85.

Elizabeth Pennisi (2012). "ENCODE Project Writes Eulogy for Junk DNA" Science 337 (6099): 1159–1160. doi:10.1126/science.337.6099.1159.

Ewen SWB, Pusztai A. (1999). Effect of diets containing genetically modified potatoes expressing *Galanthus nivalis* lectin on rat small intestine. Lancet (N Am Ed), 354:4.

Fütterer J (1995). Genetic markers and expression signals. In: Potrykus I and Spangenberg G (eds.), Gene Transfer to Plant. Springer Lab Manual, Berlin, 311-324.

Glover D. (2009).'Undying Promise: Agricultural Biotechnology's Pro-poor Narrative, Ten Years on'. STEPs Working Paper 15. Brighton: STEPs Centre; 2009.

Hammer RE, Pursel VG, Rexroad CE, Wall RJ, Palmiter RD, Brinster RL (1985). Nature 315:680-683.

Ho MW (1998) Genetic Engineering Dream or Nightmare? Gateway, Gill & Macmillan, Bath and Dublin, 1998, 1999, Chapter on Perils amid Promises of Genetically Engineered Foods.

Ho M-W (2011). The best kept secret of GM crops. Witness Statement to ACRE for ACRE open hearing on the criticisms of T25 GM maize risk assessment. Institute of Science in Society. 2002 [http://www.i-sis.org.uk/secretGMcrops.php], Accessed 20 May 2011.

Ho MW, Steinbrecher RA. (1998). Fatal flaws in food safety assessment: critique of the joint FAO/WHO Biotechnology and Food Safety Report. Environmental and Nutritional Interactions, 2:51-84.

Ho MW, Ryan A, Cummins J. (2004). Cauliflower mosaic viral promoter – a recipe for Disaster? *Microbial* IUCN Environmental Law Centre (2004) Genetically Modified Organisms and Biosafety: A background paper for decision-makers and others to assist in consideration of GMO issues. The World Conservation Union August.

Ho MW, Ryan A, Cummins J. (2000). Hazards of transgenic plants with the cauliflower mosaic viral promoter. Microbial Ecology in Health and Disease, 12:6-11.

Ho MW (1999-2001). "Transgenic pollution by horizontal gene transfer?" in ISIS Reprints on Horizontal Gene Transfer, 1999-2001, ISIS Publications, London.

Ho MW (1999-2001) "Horizontal gene transfer happens. A practical exercise in applying the precautionary principle" in *ISIS Reprints on Horizontal Gene Transfer, 1999-2001,* ISIS Publications, London.

Horvath H, Jensen L, Wong O, Kohl E, Ulrich S, Cochran J, Kannangara C, von Wettstein D. (2001). Stability of transgene expression, field performance and recombination breeding of transformed barley lines. *Theor Appl Genet.* 1-11.

ISIS Reprints *on Transgenic Instability, 1999-2001*, ISIS Publications, London.

Joint FAO/WHO Expert Consultation on Foods Derived from Biotechnology, WHO Headquarter, Geneva, September 24-28, 2001.

James C (2010). Global Status of Commercialized Biotech/GM Crops: 2010 (ISAAA Brief No.42) International Service for the Acquisition of Agribiotech Applications. Ithaca, NY; 2010.

James C. (2007). Global Status of Commercialized Biotech/GM Crops: 2007 (ISAAA Brief No. 37) International Service for the Acquisition of Agribiotech Applications. Ithaca, NY; 2007.

Jefferson RA, Kavanagh TA, Bevan MW (1987). GUS fusions: β-glucuronidase as a sensitive and versatile gene fusion marker in higher plants. EMBO J, 3901-3908.

Konig A, Cockburn A, Crevel RWR, et al. (2004). Assessment of the safety of foods derived from genetically modified (GM) crops. Food Chem Toxicol., 42:1047-88.

Kumar S, Fladung M. (2000). Transgene repeats in aspen: molecular characterisation suggests simultaneous integration of independent T-DNAs into receptive hotspots in the host genome. Mol Gen. Gent, 264:20-8.

Kumpatla SP, Chandrasekharan MB, Iyer LM, Li G, Hall TC. (1998). Genome intruder scanning and modulation systems and transgene silencing. Trends in Plant Sciences, 3:96-104.

Miller DG, Rutledge EA, Russell DW. (2002). Chromosomal effects of adeno-associated virus vector integration. Nature Genetics, 30:147-148.

Moens W, Collard JM (2002). GM plants containing antibiotic resistance genes. http://www.biosafety.be/ARGMO /GMO_plants.html (Last accessed on February 2013).

Nordlee JA (1996). "Identification of Brazil-Nut Allergen in Transgenic Soybeans''. New England Journal of Medicine, 334:688-692.

Paarlberg R. (2010). GMO foods and crops: Africa's choice. New Biotechnology.

Qayum A, Sakkhari K. (2005). Bt Cotton in Andhra Pradesh: A Three-year Assessment. The First Ever Sustained Independent Scientific Study of Bt Cotton in India. Deccan Development Society, Andhra Pradesh Coalition in Defence of Diversity, Permaculture Association of India.

Quist D, Chapela IH. (2001)Transgenic DNA introgressed into traditional maize landraces in Oaxaca, Mexico. Nature, 414:541-543.

Rauschen S (2010). A case of ''pseudoscience?" A study claiming effects of the Cry1Ab protein on larvae of the two-spotted ladybird is reminiscent of the case of the green lacewing. Transgenic Research (2010) 19:13–16 Doi 10.1007/s11248-009-9301-0

Schrott M (1995). Selectable market and reporter genes. In: Potrykus I and Spangenberg G (eds.), Gene Transfer to Plant. Springer Lab Manual, Berlin, 325-336.

RS (2003). Where is the evidence that GM foods are inherently unsafe. http://royalsociety.org http://royalsociety.org. Submission-to-the-Governments-GMScience-Review/http://royalsociety.org/GM-crops-modern-agriculture-andthe-environment—Report-of-a-Royal-Society-Discussion-Meeting-held-on-11February-2003.

Schmidt JEU, Braun CU, Whitehouse LP, Hilbeck A (2009). Effects of activated *Bt* transgene products (*Cry1Ab, Cry3Bb*) on immature stages of the ladybird *Adalia bipunctata* in laboratory ecotoxicity testing. Arch Environ Contam Toxicol 56(2):221–228.

Séralini GE, Spiroux de Vendômois J, Cellier D, et al. (2009). How sub-chronic and chronic health effects can be neglected for GMOs, pesticides or chemicals. International Journal of Biological Sciences, 5:438-443.

Séralini GE, Clair E, Mesnage R, Gress S, Degarge N, Malatesta M, Hennequin D, Vendomois JS (2012) Long term toxicity of a Roundup herbicide and a Roundup-tolerant genetically modified maize. Food Chem. Toxicol. (2012), http://dx.doi.org/10.1016/j.fct.2012.08.005. (Last accessed on February 20 2013).

Stirn S, Lorz H (2006). Genetically Modified Plants. In: Heller KJ (ed), Genetically Engineered Food–Methods and Detection. Wiley VCH, Weinheim, 28-63.

Svitashev S, Ananiev E, Pawlowski WP, Somers DA. (2000). Association of transgene integration sites with chromosome rearrangements in hexaploid oat. Theoretical and Applied Genetics, 100:872-880.

Tax FE, Vernon DM. T. (2001). DNA-associated duplication/translations in Arabidopsis. Implications for mutant analysis and functional genomics. Plant Physiology, 126:1527-38.

Vendômois JS, Cellier D, Vélot C, Clair E, Mesnage R, Séralini GE (2010). Debate on GMOs health risks after statistical findings in regulatory tests. International Journal of Biological Sciences, 6:590-598.

Vendômois JS, Roullier F, Cellier D, Séralini GE (2009) A comparison of the effects of three GM corn varieties on mammalian health. International Journal of Biological Sciences, 5:706-726.

Windels P, Taverniers I, Depicker A, Van Bockstaele E, De Loose M (2001). Characterisation of the Roundup Ready soybean insert. European Food Research and Technology. DOI 10.1007/ s002170100336, © Springer-Verlag. See also "Scrambled genome of Roundup Ready soya" by Mae-Wan Ho, ISIS Reprints on Transgenic Instability, *1999-2001*, ISIS Publications, London.

3. Strategies for the production of gluten-free bread from sorghum-cassava flour blend

Calvin Onyango[1*], Guenter Unbehend[2], Eunice A. Mewa[1], Anne W. Mutahi[1], Meinolf G. Lindhauer[2] and Michael W. Okoth[3]
[1]Food Technology Division, Kenya Industrial Research and Development Institute, Nairobi, Kenya.
[2]Institut für Sicherheit und Qualität bei Getreide, Max Rubner-Institut, Detmold, Germany.
[3]Department of Food Science, Nutrition and Technology, University of Nairobi, Nairobi, Kenya.
**Corresponding author: calonyango@yahoo.com*

INTRODUCTION

Wheat bread is one of the oldest and most popular staple foods in the world. Bread is an unstable, elastic, solid foam, the solid part of which contains a continuous phase composed in part of an elastic network of cross-linked gluten molecules and in part of leached starch polymer molecules, primarily amylose, both uncomplexed and complexed with polar lipid molecules, and a discontinuous phase of entrapped, gelatinized, swollen, deformed starch granules (Gray and Bemiller, 2003).

Wheat is the most important crop for breadmaking because of its supreme baking performance in comparison with other cereals. The unique dough-forming and breadmaking properties of wheat are attributed to gliadin and glutenin proteins which on mixing with water develop into gluten. Gluten is an artifact of processing flour and does not exist as such in the wheat kernel (Khan and Nygard, 2006). It is only formed when the two major classes of wheat protein, the gliadins and glutenins, interact to form a viscoelastic mass when they are hydrated. The glutenin fraction gives a rough rubbery mass when fully hydrated whereas the gliadin fraction forms a viscous fluid mass on hydration. Thus, dough strength and elasticity is due to the glutenin fraction, whereas dough viscosity and extensibility is due to the gliadin fraction (Wieser, 2007). The resultant gluten complex that they form is responsible for the elastic and extensible properties of dough and the protein-starch interaction that is related to gas cell formation, including stabilization and retention of gas cells within the dough structure. Gluten is also the main contributor to primary characteristics of bread such as the volume, chewy texture and flavour of the crust.

Gluten-free bread is made from non-wheat or non-gluten containing flours and starches, including cereals (such as rice, maize, sorghum and millet), pseudocereals (such as amaranth, buckwheat and quinoa), root or tuber crops (such as potatoes and sweet potatoes), plantains (cooking bananas) and legumes (such as beans and chick peas). The absence of gluten in these materials results in hydrated mixture with rheological character ranging from liquid batter to firm inelastic dough. Unlike wheat dough, gluten-free dough is unable to develop a continuous protein network with characteristics similar to gluten that can embed starch due to their different protein properties (Hüttner and Arendt, 2010). The air retained in the dough during mixing and carbon dioxide developed by yeast does not find in such an unstable system enough coherent structures to be entrapped to make the dough rise (UNECA, 1998). Part of the gases escape too early and part are retained to form irregular, unstable cells. Consequently, the baked product has deficient quality characteristics such as crumbling dry texture, low specific volume, firm crumb and short shelf-life.

To overcome the technological difficulties of making gluten-free bread, it is necessary to incorporate gluten substitute(s) in the recipe. These compounds help to create a three-dimensional gas retaining network while permitting the dough to expand in volume. A single gluten substitute does not exist and the common approach is to incorporate modified starches, modified gluten-free flours (such as that obtained by sourdough fermentation or malting), emulsifiers, enzymes, hydrocolloids, non-gluten proteins, prebiotics and combinations thereof, to gluten-free recipes. Generally, gluten substitutes improve the ability of gluten-free dough to retain carbon dioxide, strengthen dough cohesiveness and improve crumb texture, appearance and the keeping quality of gluten-free bread.

Since gluten-free bread recipes are highly variable, we will limit this review mainly to the latest research work of our research group that has focused on the development of gluten-free bread from a composite flour mixture of sorghum and cassava.

JUSTIFICATION FOR THE PRODUCTION OF GLUTEN-FREE BREAD

Wheat is the single largest food import in Kenya. Between 1990 and 2010, wheat imports have fluctuated between 130,000 and 840,000 tons/year at a cost of US$ 23-220 million/year (Chapter 1). The average wheat imports in the country between 1990 and 2010 represented about 60% of total wheat consumption. Over the last two decades, Kenya has spent cumulatively over US$ 2 billion on wheat imports. This is a staggering amount of money that could be used to improve health service provision, literacy, infrastructure and communication. It could also be used to increase access to safe drinking water, improve sanitation, and decrease social inequities and poverty in the country. Unless there is a paradigm shift in consumer desire for wheat as a status attainment staple food crop, it is reasonable to assume that future demand for wheat will continue to increase.

Sorghum and cassava are well adapted to a wider range of agro-ecological zones of Kenya than wheat. These crops can be grown with little management, have high production potential per unit of land and form part of mixed cropping systems. They are valuable assets for household food security for many farmers in times of drought. Despite these positive attributes, local consumption of sorghum and cassava is low, as compared to wheat-based products, because of the lack of variety of convenient ready-to-eat sorghum-, or cassava-based foods. The crops are mainly processed into flour which is used to make thick or thin porridge. Nonetheless, because of the rapidly increasing population and the high cost of imported wheat, it is essential to develop novel food products from these crops that can appeal to a wide cross-section of consumers.

Although wheat bread is one of the most widely consumed staple foods in the world, certain individuals have to totally avoid bread in their diet because they suffer from a medical condition referred to as coeliac disease. Coeliac disease is a life-long auto-immune mediated enteropathy which is triggered when genetically susceptible individuals ingest wheat proteins (gliadins and glutenins) and related proteins in rye, barley, durum wheat, spelt wheat, kamut, einkorn and triticale. When these proteins enter the digestive tract, they are broken down into peptides which trigger the immune response. The disease is manifested as damage to the mucous membrane of the small intestine (duodenum and jejunum) leading to malabsorption of basic nutrients like proteins, carbohydrate and fat; micronutrients such as vitamins and minerals; and in some cases water and bile salts. The individual also develops associated disorders that can affect almost any organ or body system. The classic intestinal symptoms of the disease include dental enamel defects, abdominal distension, constipation, diarrhoea and vomiting. Extra-intestinal manifestations of the disease include iron-deficiency anaemia, weight loss, infertility, arthritis, osteoporosis and dermatitis herpetiformis. The individual may also develop neurological and psychological disturbances (Catassi and Fasano, 2009; Green, 2009).

Another adverse immune reaction associated with ingestion of wheat-containing food by certain individuals is wheat allergy. Wheat allergy is an immunological reaction that is mediated by immunoglobulin E. Among the wide spectrum of gluten antigens that elicit the allergic reaction are the α-, β-, γ- and ω-gliadins and low molecular weight glutenins. The allergic reactions are characterized by skin reactions (such as urticaria and angioedema), gastrointestinal reactions (such as vomiting and diarrhoea), nausea, rhinitis and asthma (Catassi and Fasano, 2009).

The only effective treatment for coeliac disease and wheat allergy is based on life-long exclusion of gluten-containing cereals, whether as ingredients or additives, from the diet (Box 1). Strict adherence to a gluten-free diet prevents morbidity and reduces the incidence of the associated gastrointestinal malignancy for coelic patients. This means that, with the exception of wheat, rye and barley, all cereals in the Poaceae (formerly Gramineae) family are suitable for the production of gluten-free bread. Pseudocereals (amaranth, buckwheat, quinoa), root and tuber crops (such as sweet potatoes, potatoes, yams, arrow roots) and plantains are also carbohydrate-rich gluten-free foods that can be tolerated by people suffering from coeliac disease and wheat allergy.

BOX 1: GLUTEN-FREE FOODS

Gluten-free foods are dietary foods

consisting of or made only from one or more ingredients which do not contain wheat (i.e. all *Triticum* species, such as durum wheat, spelt and kamut), rye, barley, oats[1] or their crossbred varieties, and the gluten level does not exceed 20 mg/kg in total (i.e. 20 ppm), based on the food as sold or distributed to the consumer, and / or

consisting of one or more ingredients from wheat (i.e. all *Triticum* species, such as durum wheat, spelt, and kamut), rye, barley, oats[1] or their crossbred varieties which have been specially processed to remove gluten, and the gluten level does not exceed 20 mg/kg in total (i.e. 20 ppm), based on the food as sold or distributed to the consumer.

[1]Oats can be tolerated by most but not all people who are intolerant to gluten. Therefore, the allowance of oats that are not contaminated with wheat, rye or barley in foods covered by this standard may be determined at the national level.

Source: CODEX STAN 118-1979 (2012).

SORGHUM

Sorghum is a monocotyledonous plant that belongs to the Poaceae (formerly Gramineae) family, Panicoideae subfamily and Andropogeneae tribe. Two of the best known species are *Sorghum vulgare* and *Sorghum bicolor* (L.) Moench. Due to genetic diversity, the grains vary widely in shape, size and colour. The kernels are typically round, varying in weight from 20-30 g per 1000 kernels (Delcour and Hoseney, 2010). The kernel is a naked caryopsis, typically 2-5 mm in length and 2-3 mm thick at the widest point (Taylor and Belton, 2002). The colour of the kernel varies from white or yellow to red, whereas the endosperm colour can be yellow or white (Schober and Bean, 2008).

Sorghum grain is made up of a pericarp, endosperm (storage tissue) and germ (embryo). The pericarp region comprises a pericarp, testa (seed coat) and aleurone layer. Sorghum is unique in that it is the only cereal grain that has starch granules in the pericarp. The testa separates the pericarp from the aleurone layer. The testa is thin in low tannin sorghum varieties but thicker and highly pigmented in high tannin sorghums (Taylor and Belton, 2002). The outer edge of the endosperm is composed of the aleurone layer containing lipids, enzymes and protein bodies. Under the aleurone layer is the outer corneous (i.e. hard or vitreous) endosperm fraction surrounding an inner floury (i.e. soft) core (Schober and Bean, 2008).

The main nutrient component of sorghum is carbohydrate, which exists predominantly as starch (50-75%) and is located in the endosperm and pericarp of the grain. Starch granules in sorghum range from 2-30 μm in diameter and contain 20-30% amylose (Taylor and Belton, 2002). Proteins make up 9-14% of the grain (Waniska et al., 2004) and are principally located in the endosperm. The protein quality of sorghum is poor because of the low content of essential amino acids such as lysine, tryptophan and threonine (Badi et al., 1990). Minor components of sorghum grain are lipids, vitamins and minerals. Although these compounds occur in small amounts, they have great influence on the processing parameters and nutrient quality of the grain. The lipid content of sorghum varies from 2.1-6.6%. The lipids are mainly located in the germ, although there are smaller amounts in the endosperm (Taylor and Belton, 2002). The whole grains are a good source of the B vitamins: thiamine (vitamin B_1), riboflavin (vitamin B_2), niacin (vitamin B_3), pyridoxine (vitamin B_6) and biotin (vitamin B_7). The main minerals in sorghum are potassium and phosphorous. Minerals and vitamins are naturally present in higher amounts in whole milled flour than in refined flour but they can be added back to refined flour through the process of enrichment.

Sorghums are rich in phytochemicals (i.e. polyphenols, anthocyanins, phytosterols and policosanols) with potential significant impact on human health. The ability of sorghum tannins to complex with proteins has the potential to slow digestion and reduce caloric intake of foods. Tannin-containing sorghum varieties show high antioxidant activities against different free radicals *in vitro* that compare favourably with high antioxidant fruits and vegetables. Sorghum bran is rich in phytosterols and policosanols that promote cardiovascular health. There is epidemiological evidence to suggest that sorghum consumption reduces the risk of certain types of cancers (Rooney and Awika, 2005; Awika and Rooney, 2004).

CASSAVA

Cassava is a dicotyledonous perennial plant that belongs to the Euphorbiaceae family. The root is rich in starch and is a valuable source of energy in human nutrition. Root size, shape and colour depend on the variety and environmental conditions. Cassava roots are about 15-100 cm long and 3-15 cm wide. They are cylindrical, conical or oval, with a coffee, pink, or cream-coloured peel that is covered by a thin brown bark (the periderm). The parenchyma (the edible portion of the root) is white, cream or yellow. Cassava plants produce 5-10 roots weighing 0.5-2.5 kg each (Wheatley et al., 2003).

The major nutrient component of cassava is carbohydrate which makes up over 90% of parenchyma dry-weight. Plant carbohydrates include celluloses, gums and starches, but starches are the main source of nutritive energy because celluloses are not digested (FAO, 1990). The starch content in cassava is 64-72% on dry-weight-basis. The other carbohydrate fractions are sugar and non-starch polyscacchrides such as celluloses and hemicelluloses. The protein, fat and ash contents of cassava are generally below 1%. The fibre content is more variable and increases with plant age. Cassava roots contain cyanogenic glycoside which is converted to toxic hydrocyanic acid when it comes into contact with linamarase, an enzyme that is released when the cells of cassava roots are ruptured. The cyanogenic glycoside content of cassava roots varies widely and is related to environment factors, size of roots and moisture content. The values range from 30-100 mg/kg in low-cyanogenic cultivars for direct consumption to 1350 mg/kg in industrial varieties used for processing (Wheatley et al., 2003).

GLUTEN-FREE BREAD FROM SORGHUM AND CASSAVA

Sorghum and cassava flours are unable to form cohesive, gas-holding viscoelastic dough when added to water because they do not contain gluten. The rheological character of the mixture ranges from inelastic dough to free-flowing batter, depending on the nature and amount of ingredients and additives. Onyango et al. (2011a) reported that the consistency of sorghum-starch gluten-free dough changes from a stiff inelastic mass to a thin pourable batter with increasing starch content. At a starch to sorghum ratio of 10:90 the mixture has the rheological character of inelastic dough that can be shaped but cannot stretch. On the other hand, the mixture has the rheological character of thin pourable batter when the sorghum content is decreased relative to starch content. By contrast, when pregelatinised starch is used, dough strength increases with increasing content of pregelatinised starch (Onyango et al., 2011b). The rheological character of gluten-free mixture is also affected by the amount of mechanically damaged starch in the flour (Schober and Bean, 2008). Flour with a higher amount of damaged starch will produce thicker batter than flour with a lower amount of damaged starch.

Starch is the most important structure-forming ingredient in gluten-free bread due to the absence of gluten that performs this role in wheat bread. The overall bread quality is influenced by the interaction of starch with the other dough components and its behaviour on hydration, heating and cooling. Increasing starch content in gluten-free recipes induces early onset gelatinization, resulting to an early increase in crumb consistency during baking. This facilitates the development of a cohesive crumb network that traps gas bubbles and prevents loss of carbon dioxide and crust collapse. Increasing starch content also improves the volume of gluten-free bread by diluting the endosperm and bran particles which interfere with the stability of the sorghum gel and liquid films around the gas cells (Taylor et al., 2006). Increasing starch content in gluten-free recipes also improves crumb texture properties. Onyango et al. (2011a) investigated the effect of different starch concentrations on the crumb properties of sorghum-based gluten-free bread and found that crumb firmness and chewiness decreases with increasing starch content, whereas springiness, cohesiveness and resilience increases. However, starch breads have certain disadvantages that include rapid staling, high glycemic index and bland flavour. Furthermore, starch breads are poor in micronutrients, proteins and dietary fibre (Schober, 2009). Figure 1 shows an outline for the production of gluten-free bread from sorghum flour and cassava starch at a ratio of 1:1. Figure 2 shows the baking pan with the lid and the appearance of the gluten-free breads.

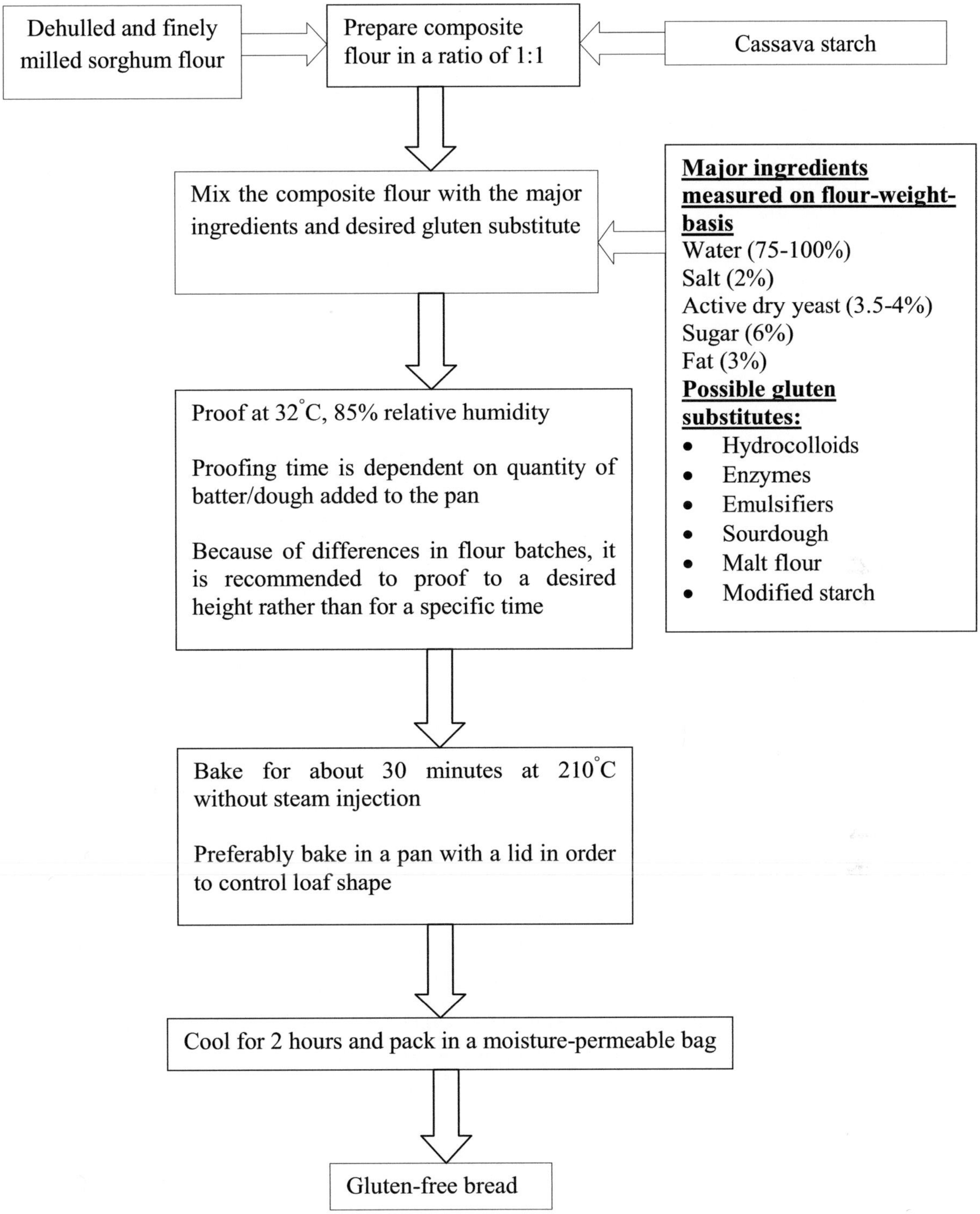

Figure 1. Outline of steps required to make gluten-free bread from sorghum flour and cassava starch.

Figure 2. Gluten-free bread prepared from sorghum flour and cassava starch (b-d) at a ratio of 1:1 and 100% water (weighed on flour-weight-basis). The breads were baked in pans with lids (a-b) to control loaf shape.

Starches from different botanical origins have variable sizes and shapes and possess diverse ranges of pasting, gelling, thermal and textural characteristics, depending on their composition and structural properties (Abdel-Aal, 2009). Onyango et al. (2011a) studied the rheological character and crumb quality of gluten-free batter and bread, respectively, prepared from blends of sorghum flour and either cassava, maize, rice or potato starches. They reported that batter prepared from sorghum flour and rice starch was less firm than batter from sorghum flour and cassava, maize or potato starches. On baking, they found that the crumb texture of sorghum-cassava bread was softer, springier and more resilient and cohesive than the crumb texture of bread prepared from sorghum with either maize, potato or rice starch.

It has previously been argued (Taylor and Belton, 2002; Hugo et al., 1997; Olatunji et al., 1992b) that when pregelatinised cassava starch is blended with or substituted for native starch, the crumb quality of gluten-free bread is improved. The rheological character of batter prepared from sorghum flour and pregelatinised cassava starch is distinctly different from that prepared from sorghum flour and native cassava starch (Onyango et al., 2011b). Batter prepared from sorghum flour and pregelatinised cassava starch has firm, inelastic character and can be shaped because pregelatinised starch, which is rich in damaged starch, has several exposed hydroxyl groups that form multiple bonds with water. However, formation of a firm inelastic mass from sorghum and pregelatinised starch does not translate in gluten-free bread with better crumb properties. Bread prepared from sorghum flour and pregelatinised starch is firm and has a wet and sticky (i.e.adhesive) crumb (Onyango et al., 2011b), which is due to the high water-binding capacity of pregelatinised starch. On the other hand, gluten-free bread made from sorghum flour and native cassava starch has a soft non-adhesive crumb (Figure 3).

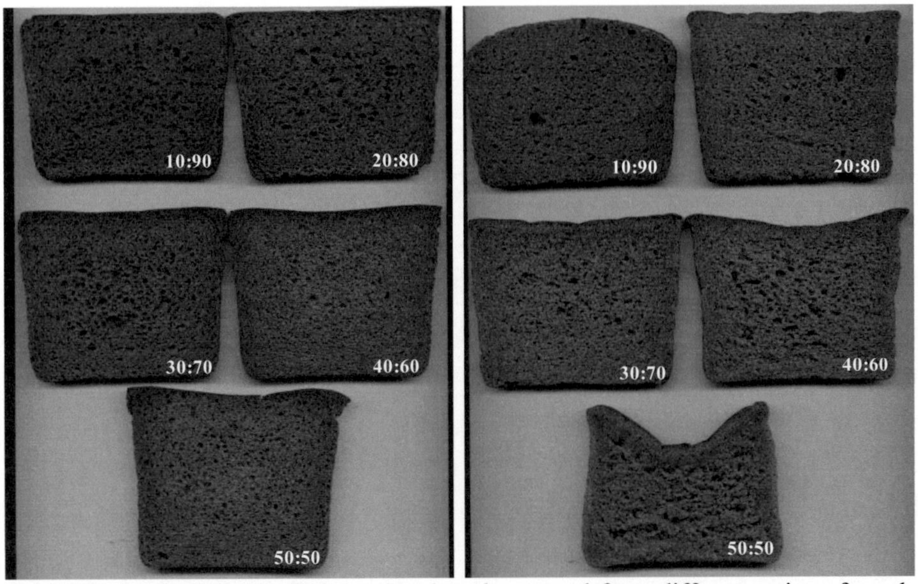

Figure 3. Cross-sections of gluten-free bread prepared from different ratios of sorghum flour and native cassava starch (left) and sorghum flour and pregelatinised cassava starch (right). Source: Onyango et al. (2011a).

The kernels of sorghum varieties show tremendous diversity in terms of colour, weight and composition (Table 1). These differences, in turn, affect the processing properties and taste, texture and nutritional value of sorghum-based products. Schober et al. (2005) compared the effect of nine sorghum cultivars and commercial sorghum flour on the quality of gluten-free bread (70 parts sorghum and 30 parts corn starch). They found that crumb grain (pore size and number of pores) and firmness were affected by sorghum hybrid and concluded that the amount of mechanically damaged starch in the flour, which was influenced by kernel hardness, was responsible for the differences. Varieties with higher amounts of damaged starch had coarser crumbs (because of damaged starch's greater susceptibility to the action of amylases) than those with lower amounts of damaged starch. Sorghum variety, however, did not affect bread volume, height, bake loss and water activity.

Table 1. Composition of five sorghum varieties in the Kenyan market.

Sorghum variety	Pericarp colour	1000-kernel weight	Carbohydrate	Starch	Protein	Fat	Fibre	Ash
Gadam	White	17.87	79.63	64.41	13.19	3.62	2.00	1.56
Kaguru	Pink	14.58	83.05	60.69	10.51	2.77	2.22	1.46
KARI Mtama II	Reddish-brown	21.29	83.12	61.81	9.48	2.77	2.61	2.02
Seredo	Reddish-brown	24.96	80.97	58.77	11.46	3.37	2.49	1.71
Serena	Brown	22.15	84.76	63.49	8.16	3.23	2.24	1.61

All numeric values are given as g/100 g dry-weight basis.

The choice of non-gluten protein in a gluten-free recipe is determined by its availability, cost, ease of production, types and levels of inhibitors, allergenic potential, and nutritional and functional influence on product characteristics. For instance, whereas milk sugar (lactose) causes bloating and diarrhoea in lactose-intolerant people, it also supplements the limiting amino acids (lysine, methionine and tryptophan) and improves the flavour and crust colour of gluten-free bread. In another example, Ribotta et al. (2004) reported that non-heat treated soybean flour gives gluten-free bread with better quality (i.e. well-aerated bread structure and a good loaf volume) than heat-treated soybean flour. They argued that heat treatment inactivates soybean enzymes and causes protein aggregation and subsequent loss of protein solubility, which have negative effects on the application of soybean flour in gluten-free bread. Onyango et al. (2009a) compared the effect of egg white, skim milk powder, soy protein isolate and soy protein concentrate on the crumb quality of gluten-free bread prepared from sorghum and pregelatinised cassava starch. They reported crumb firmness of 1,000, 1,573, 2,249 and 2,168 g (force) for gluten-free bread containing egg white, skim milk powder, soy protein isolate and soy protein concentrate, respectively. The corresponding values for specific volume were 3.26, 2.91, 2.69 and 2.77 cm^3/g, whereas the staling rates were 58, 260, 168 and 229 g/day. Figure 4 shows the effect of different amounts of egg white on the volume and appearance of gluten-free bread prepared from sorghum and pregelatinised cassava starch. The volume and shape of the bread improved with increasing content of egg white in the formulation. The superior functional property of egg white powder is associated with its capacity to create continuous web-like structures in gluten-free dough and bread, which resembles the network structure in wheat dough and bread (Ahlborn et al., 2005; Moore et al., 2004).

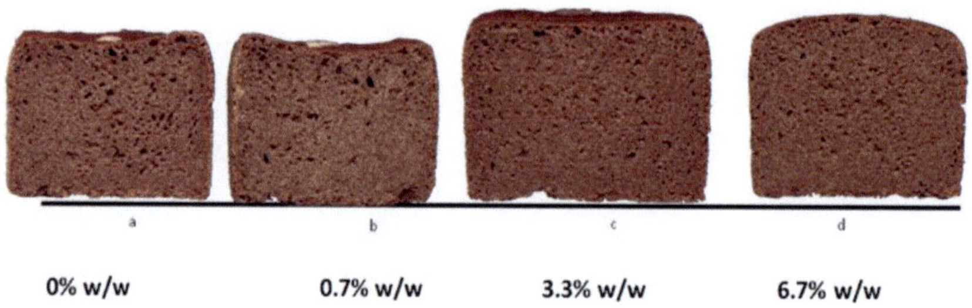

Figure 4. Effect of different amounts of egg white on volume of gluten-free bread prepared from sorghum and pregelatinised cassava starch.

Hydrocolloids (also commonly called gums) are substances consisting of hydrophilic long-chain, high molecular weight molecules, usually with colloidal properties, that in water-based systems produce gels or thickened liquids (Hoefler, 2004). They include a wide range of polysaccharide and protein biopolymers that are derived from natural sources such as plants, animals, sea weeds and microorganisms, but can also be of synthetic origin. Hydrocolloids are naturally present or added to a wide range of foodstuffs to modify texture, appearance as well as improve product stability (Abdel-Aal, 2009). The functional effects of hydrocolloids in baked products stem from their ability to modify dough handling properties, increase quality of fresh bread and extend the shelf-life of stored bread (Rosell et al., 2001; Guarda et al., 2004).

Pentosans are naturally occurring hydrocolloids in sorghum. They are found in small amounts in the pericarp (ca. 1.5%) but because most are lost during refining, they have minimal effect on crumb properties of sorghum bread. Nonetheless, water-extractable pentosans can be used as additives to improve volume and retard staling of gluten-free sorghum bread (Schober and Bean, 2008). They contribute to the water-holding capacity and viscosity of dough and produce a film inside the vacuoles of the fermenting dough that contributes to gas retention and loaf volume (Abdel-Aal, 2009).

Hydrocolloids have been used in gluten-free bakery formulations to mimic the viscoelastic properties of gluten and improve bread quality (Crockett et al., 2011; Hüttner and Arendt, 2010; Lazaridou et al., 2007). They help to stabilize and retain gas cells in the dough, promote water absorption, improve bread volume, height and crumb porosity and retard crumb staling (Gallagher et al., 2004; Bemiller, 2008). The specific effect of hydrocolloids is dependent on its origin, chemical structure, concentration, technique of application and properties of the flour. Onyango et al. (2009) studied the effect of five cellulose-derivatives on the crumb texture of gluten-free bread prepared from sorghum and pregelatinised cassava starch and found that firmness decreased in the following order: hydroxypropylcellulose > microcrystalline cellulose > hydroxypropylmethylcellulose (HPMC) > methylcellulose > carboxymethylcellulose. Anton and Artfield (2007) reviewed the application of hydrocolloids in gluten-free recipes and concluded that HPMC and xanthan gum were the most effective in mimicking gluten properties in the dough and final product. HPMC is especially effective in gluten-free recipes because of its ability to thicken dough (thereby preventing phase separation) and stabilize foams (thereby improving gas retention, crumb structure and loaf volume) (Schober, 2009; Crockett et al., 2011, Sivaramakrishnan et al., 2004). Xanthan gum, on the other hand, thickens dough but does not stabilize foams. Consequently, gluten-free bread containing HPMC has a more regular crumb, larger volume and slower staling rate than that containing xanthan gum or without any additive (Schober and Bean, 2008; Schober, 2009).

Enzymes are widely applied as food processing aids because they are natural, non-toxic food additives that are preferred by consumers over their chemical counterparts (James and Simpson, 1996). The effect of a number of enzymes on the quality of sorghum-based gluten-free bread has been investigated. Renzetti and Arendt (2009) found that the bread making performance of sorghum flour is not improved when its proteins are degraded by proteases but is enhanced when glucose oxidase is added. Olatunji et al. (1992a) reported that fungal amylase improves the sensory characteristics of sorghum-based gluten-free bread, whereas Schober et al. (2007) attributed limited α-amylase activity in sourdough fermented sorghum bread to the acidic environment which inactivated the enzyme. Onyango et al. (2010a) reported that α-amylase increased crumb adhesiveness but decreased firmness, cohesiveness, springiness, resilience and chewiness of sorghum-bread supplemented with native or pregelatinised cassava starch. They also noted that the amount of α-amylase in the gluten-free formulation and the nature of starch (native or pregelatinised) influenced the crumb texture. The recipe that had pregelatinised cassava starch was more susceptible to enzymatic degradation which contributed to improved crumb cohesiveness and resilience. However, this bread also had an undesirably wet and sticky crumb (i.e. high adhesiveness), which could be attributed to enzyme-mediated generation of high levels of low molecular weight branched dextrins (degree of polymerization 10-100) in the bread crumb (Carroll et al., 1987).

Transglutaminase (protein-glutamine γ-glutamyltransferase, EC 2.3.2.13) is an enzyme that promotes covalent crosslinking between proteins through the reaction between ε-amino group on protein-bound lysine residues and a γ-carboxyamide group on protein-bound glutamine residues. It can also catalyse the introduction of free amine groups into proteins through the amide moiety of a glutamine residue and, in the absence of available amines, hydrolyse glutamine residues to glutamate residues (i.e. deamidate glutamine) (Gerrard and Sutton, 2005). Transglutamiases are present in plant, animal and microbial cells but only microbial transglutaminase (mTG) is used in the food industry to modify protein functionality (Dekking et al., 2008). Onyango et al. (2010b) reported that mTG increased the elastic properties of gluten-free dough and crumb firmness and chewiness of gluten-free bread prepared from sorghum and pregelatinised cassava starch. They attributed these changes to increased molecular weight of protein fractions in the gluten-free recipe due to the cross-linking action induced by mTG. Similar effects have been observed in bread made from rice (Gujral and Rosell, 2004), buckwheat, brown rice or maize flour (Renzetti et al., 2008). Despite the positive functional role of mTG in gluten-free recipes, it should be noted that this enzyme is not recommended for processing cereals that are meant for consumption by coeliacs because it deamidates proteins in dough to generate the epitope that activates T cells in the coeliac response (Dekking et al., 2008).

Emulsifiers (surface-active agents or surfactants) are substances that lower the interfacial energy between two immiscible phases thus facilitating the dispersion of one phase into the other. Emulsifiers are commonly used in bakery products to strengthen the dough and soften the crumb. The level of natural emulsifiers in sorghum is low and so they have to be added to gluten-free dough to modify its rheological properties and the subsequent bread crumb texture. Onyango et al. (2009) studied the effect of four emulsifiers on the crumb texture of gluten-free bread prepared from sorghum and pregelatinised cassava starch. They reported that crumb firmness decreased in the following order: DATEM > calcium stearoyl-2-lactylate > glycerol monostearate > sodium stearoyl-2-lactylate. They also noted that although emulsifiers soften the crumb texture, they also create the undesirable crumb property of increased crumbliness (i.e. they decrease crumb cohesiveness).

Sourdough is a mixture of flour and water and other ingredients that is fermented by naturally occurring lactic acid bacteria and yeasts (Gobbetti et al., 2008). Other than being natural and additive-free, other positive attributes associated with sourdough in baking are improved texture, flavour, nutritional value and shelf-life (Gobbetti et al., 2008). These characteristic features derive from the complex metabolic activities (such as acidification, production of exopolysaccharides, proteolytic- amylolytic- and phytase activity, and production of antimicrobial substances) of the sourdough microflora (Moroni et al., 2009). Schober et al. (2007) prepared gluten-free sorghum bread (70 parts sorghum and 30 parts potato starch) from sorghum flour that had been subjected to sourdough fermentation. They found that the bread developed a continuous cohesive crumb when sorghum proteins were degraded into peptides during sourdough fermentation. They attributed the improved crumb structure to the inability of the newly formed peptides to interfere with the starch gel. They also found that sourdough fermentation improved the shape of the bread and decreased the staling rate. Although sourdough gluten-free bread may be objectionable to consumers not accustomed to sour tasting bread, the sour taste can be neutralized with calcium carbonate (Schober et al., 2007).

CONCLUSION

Although good quality bread can be made from wheat, economic and health reasons have necessitated the development of gluten-free bread. Gluten-free bread is made from non-wheat or non-gluten containing flours and starches such as sorghum flour and cassava starch. Gluten-free dough has poor viscoleastic and gas retaining properties, which result in gluten-free bread with low specific volume, and poor crumb texture and keeping quality. The quality of gluten-free bread can be improved significantly by adding gluten-substitutes such as hydrocolloids, emulsifiers, enzymes, and modified flours and starches. These substances mimic the viscoelastic behaviour of gluten, and thus serve to improve the quality of gluten-free bread.

REFERENCES

Abdel-Aal E.-SM. (2009). Functionality of starches and hydrocolloids in gluten-free foods. In: Gluten-free Food Science and Technology (ed. Gallagher, E.) Wiley-Blackwell, Oxford. Pp. 200-224.

Ahlborn GJ, Pike OA, Hendrix SB, Hess WM, Huber CS. (2005). Sensory, mechanical, and microscopic evaluation of staling in low-protein and gluten-free breads. Cereal Chemistry, 82:328-335.

Anglani C. (1998). Sorghum for human food – A review. Plant Foods for Human Nutrition, 52, 85-95.

Anton AA, Artfield SD. (2007). Hydrocolloids in gluten-free breads: A review. International Journal of Food Sciences and Nutrition, 59:11-23.

Bemiller JN. (2008). Hydrocolloids. In: Gluten-free Cereal Products and Beverages (ed. Arendt, E. K. and Bello, F. D.). Academic Press, London. Pp. 203-214.

Carroll JO, Boyce COL, Wong TM, Starace CA. (1987). Bread antistaling method. United States Patent 4,654,216.

Catassi C, Fasano A. (2008). Celiac disease. In: Gluten-free Cereal Products and Beverages (ed. Arendt, E. K. and Bello, F. D.). Academic Press, London. Pp. 1-27.

CODEX STAN 118–1979 (2012). Codex standard for foods for special dietary use for persons intolerant to gluten. CODEX STAN 118–1979. Available online at http://www.codexalimentarius.org/standards/list-of-standards/en/?provide=standards&orderField=fullReference&sort=asc&num1=CODEX (accessed on 02 November 2012).

Crockett R, Ie P, Vodovotz Y. (2011). How do xanthan and hydroxypropylymethylcellulose individually affect physichochemical properties in a model gluten-free dough? Journal of Food Science, 76, E274-E282.

Cureton P, Fasano A. (2009). The increasing incidence of celiac disease and the range of gluten-free products in the market place. In: Gluten-free Food Science and Technology, (ed. Gallagher, E.). Wiley-Blackwell, UK. Pp. 1-15.

Delcour JA, Hoseney RC. (2010). Principles of Cereal Science and Technology. Chapter 1, 8. Minnesota, USA: AACC International.

Dekking EHA, Van Veelen PA, de Ru A, Kooy-Winkelaar EMC, Gröneveld T, Nieuwenhuizen WF, Koning F. (2008). Microbial transglutaminases generate T cell stimulatory epitopes in celiac disease. Journal of Cereal Science, 47:339-346.

Dewettinck K, Van Bockstaele F, Kuhne B, Van de Walle D, Courtens TM, Gellynck X. (2008). Nutritional value of bread: Influence of processing, food interaction and consumer perception. Journal of Cereal Science, 48:243-257.

Gallagher E, Gormley TR, Arendt EK. (2004). Recent advances in the formulation of gluten-free cereal-based products. Trends in Food Science and Technology, 15:143-152.

Gerrard JA, Sutton KH. (2005). Addition of transglutaminase to cereal products may generate the epitope responsible for coeliac disease. Trends in Food Science and Technology, 16:510-512.

Gobbetti M, De Angelis M, Di Cagno R, Rizzello CG. (2008). Sourdough / lactic acid bacteria. In: Gluten-free Cereal Products and Beverages, (ed. Arendt, E. K. and Dal Bello, F.). Academic Press, London, Pp. 267-288.

Goesaert H, Slade L, Levine H, Delcour JA. (2009). Amylases and bread firming – an integrated review. Journal of Cereal Science, 50: 345-352.

Gray JA, Bemiller JN. (2003). Bread staling: molecular basis and control. Comprehensive Reviews in Food Science and Food Safety, 2:1-21.

Green PHR. (2009). The clinical presentation and diagnosis of celiac disease. In: Gluten-free Food Science and Technology, (ed. Gallagher, E.). Wiley-Blackwell, UK. Pp. 16-27.

Guarda A, Rosell CM, Benedito C, Galotto MJ. (2004). Different hydrocolloids as bread improvers and antistaling agents. Food Hydrocolloids, 18:241-247.

Gujral SH, Rosell MC. (2004). Functionality of rice flour modified with a microbial transglutaminase. Journal of Cereal Science, 39:225-230.

Hoefler AC. (2004). Hydrocolloids. Egan Press Handbook. AACC International, St Paul, Minnesota, Pp. 1-42.

Hugo, LF, Waniska RD, Rooney LW. (1997). Production of bread from composite flours. In Proceedings: Harnessing cereal science and technology for sustainable development (pp 100-114). Pretoria: CSIR ICC-SA Symposium.

Hüttner EK, Arendt EK. (2010). Recent advances in gluten-free baking and the current status of oats. Trends in Food Science and Technology, 21:303-312.

James J, Simpson BK. (1996). Applications of enzymes in food processing. Critical Reviews in Food Science and Nutrition, 36:437-463.

Khan K, Nygard G. (2006). Gluten. In: Bakery Products Science and Technology, (ed. Hui, Y. H., Corke, H., de Leyn, I., Nip, W-K and Cross, N.). Blackwell Publishing, Professional, USA. Pp. 97-107.

Lazaridou A, Duta D, Papageorgiou M, Belc N, Biliaderis CG. (2007). Effects of hydrocolloids on dough rheology and bread quality parameters in gluten-free formulations. Journal of Food Engineering, 79, 1033-1047.

Moore MM, Schober TJ, Dockery P, Arendt EK. (2004). Textural comparisons of gluten-free and wheat-based doughs, batters, and breads. Cereal Chemistry, 81:567-575.

Moroni A, Dal Bello F, Arendt EK. (2009). Sourdough in gluten-free breadmaking: An ancient technology to solve a novel issue? Food Microbiology, 26:676-684.

Olatunji, O., Koleoso, O. A. and Oniwinde, A. B. (1992a). Recent experience on the milling of sorghum, millet, and maize for making nonwheat bread, cake, and sausage in Nigeria. In: Utilization of sorghum and millets (ed. Gomez, M. I., House, L. R., Rooney, L. W. and Dendy, D. A. V). International Crops Research Institute for the Semi-Arid Tropics, India. Pp. 83-88.

Olatunji O, Osibanjo A, Bamiro E, Ojo O, Bureng P. (1992b). Improvement of the quality of non-wheat composite bread. In: Proceedings of fifth quadrennial symposium on sorghum and millets (pp 45-54). Schwechat: International Association of Cereal Science and Technology.

Onyango C, Mutungi C, Unbehend G, Lindhauer MG. (2009). Effect of cellulose-derivatives and emulsifiers on creep-recovery and crumb properties of gluten-free bread prepared from sorghum and gelatinised cassava starch. Food Research International, 42:949-955.

Onyango C, Mutungi C, Unbehend G, Lindhauer MG. (2010a). Batter rheology and bread texture of sorghum-based gluten-free formulations modified with native or pregelatinised cassava starch and α-amylase. International Journal of Food Science and Technology, 45:1228-1235.

Onyango, C., Mutungi, C., Unbehend, G. and Lindhauer, M. G. (2010b). Rheological and baking characteristics of batter and bread prepared from pregelatinised cassava starch and sorghum and modified using microbial transglutaminase. Journal of Food Engineering, 97:465-470.

Onyango C, Mutungi C, Unbehend G, Lindhauer MG. (2011a). Modification of gluten-free sorghum batter and bread using maize, potato, cassava or rice starch. LWT – Food Science and Technology, 44:681-686.

Onyango C, Mutungi C, Unbehend G, Lindhauer MG. (2011b). Rheological and textural properties of sorghum-based formulations modified with variable amounts of native or pregelatinised cassava starch. LWT – Food Science and Technology, 44, 687-693.

Renzetti S, Arendt EK. (2009). Effects of oxidase and protease treatments on the breadmaking functionality of a range of gluten-free flours. European Food Research and Technology, 229, 307-317.

Renzetti S, Courtin CM, Delcour JA, Arendt EK. (2008). Microstructure, fundamental rheology and baking characteristics of batters and breads from gluten-free flours treated with a microbial transglutaminase. Journal of Cereal Science, 48:33-45.

Ribotta PD, Ausar SF, Morcillo MH, Pérez GT, Beltramo DM, León AE. (2004). Production of gluten-free bread using soybean flour. Journal of the Science of Food and Agriculture, 84:1969-1974.

Rooney LW, Awika JM. (2005). Speciality sorghums for healthful foods. In: Speciality Grains for Food and Feed, (ed. Abdel-Aal, E, and Wood, P). American Association of Cereal Chemists, AACC, USA. Pp 283-312.

Rosell CM, Rojas JA, Benedito de Barber C. (2001). Influence of hydrocolloids on dough rheology and bread quality. Food Hydrocolloids, 15:75-81.

Schober TJ. (2009). Manufacture of gluten-free specialty breads and confectionery products. In: Gluten-free Food Science and Technology, (ed. Gallagher, E.). Wiley-Blackwell, UK. Pp. 130-180.

Schober TJ, Bean SR. (2008). Sorghum and maize. In: Gluten-free Cereal Products and Beverages (ed. Arendt, E. K. and Dal Bello, F.). Academic Press, London. Pp. 101-118.
Schober TJ, Bean SR, Boyle DL. (2007). Gluten-free sorghum bread improved by sour dough fermentation: biochemical, rheological and microstructural background. Journal of Agricultural and Food Chemistry, 55:5137-5146.
Schober TJ, Messerschmidt M, Bean SR. Park S-H, Arendt EK. (2005). Gluten-free bread from sorghum: quality differences among hybrids. Cereal Chemistry, 82:394-404.
Sivaramakrishnan HP, Senge B, Chattopadhyay PK. (2004). Rheological properties of rice dough for making rice bread. Journal of Food Enginnering, 62:37-45.
UNECA (1998). Technical handbook on composite flours. Processing of tropical flours into bread. United Nations Economic Commission for Africa, Addis Ababa.
Wieser H. (2007). Chemistry of gluten proteins. Food Microbiology, 24:115-119.

4. The current status of potato value chain in Kenya

George O. Abong'[1*] and Jackson N. Kabira[2]
[1]Department of Food Science, Nutrition and Technology, University of Nairobi, Nairobi, Kenya.
[2]National Potato Research Centre (KARI), Tigoni, Kenya.
[*]*Corresponding author: Email: georkoyo@yahoo.com*

SUMMARY

Potato plays a major economic role in the world and has been considered a major staple in developing countries such as Kenya where it ranks second to maize in terms of utilization. The importance of potato in Kenya is anchored on its role in alleviating poverty and fighting hunger as well as income generation thereby playing a dual role as food and cash crop. The crop is traded both in fresh and processed forms and the value chain is a source of sustainable livelihoods. Potatoes are grown by over 800,000 farmers on 160,000 hectrares and are valued at Kshs 46 billion per annum at the consumer level. Most of the varieties grown are locally adapted and are used for domestic consumption and processing.

The Kenyan potato value chain indirectly employs 2.5 million people including researchers, extension agents, seed inspectors, transporters, market agents and processors among others. They are, however faced with a number of challenges. Although the legal limit for ware potato bags is 110 kg, middle men use extended bags often over 150 kg to fleece the farmers. Due to lack of storage facilities, farmers often sell at low prices at harvest while consumers pay high prices 2-3 months later due to low supply of potatoes before the next harvest. Potato is now considered by the Ministry of Agriculture, Livestock and Fisheries as a possible alternative crop to maize which is having problems with incurable diseases from unknown sources. Many farmers in maize growing areas are now adopting potatoes for food and income security reasons. Majority of the households in potato growing zones and a number of urban dwellers use the tubers as important accompaniment to diverse dishes. On the other hand, potato crisps and French fries (chips) are important processed products that form greater parts of menus in restaurants and hotels in major urban centers with crisps assuming large share in the supermarkets, kiosks and roadside shops. There exist a great potential in the potato value chain as much as there are challenges and opportunities. This paper is a step-by-step diagnosis of the potato value chain in Kenya, from production, post-harvest handling, processing and marketing including challenges and opportunities.

INTRODUCTION

Potato plays a major economic role in the world and has been considered as a major staple in developing countries such as Kenya where it ranks second to maize in terms of utilization (Nema et al., 2008; MoA, 2009; FAOSTAT, 2011). The importance of potato in Kenya is anchored on its role in alleviating poverty and hunger as well as income generation thereby playing a dual role as food and cash crop (Gildemacher et al., 2009). Production of the crop in the country is mainly carried out in higher altitudes of between 1,500 m and 3,000 m above sea level. In these areas, the crop has generally higher yields compared to major food crops grown in the country such as maize and other cereals and pulses due to the fact that it matures fast and can be grown twice or thrice in a year (MoA, 2009). Production which is mainly carried out on small-holder farms (2 hectares on average) is confined to central, Rift Valley and eastern highland regions of the country. Small quantities of potato are, however, produced in higher altitudes of coastal and western regions of the country where the climate is suitable. The potato value chain from farm to table provides unique opportunities and challenges. There are over 20 varieties of potato in Kenya including but not limited to Tigoni, Asante, Cangi, Kenya Karibu, Kenya Sifa, Kenya Mpya, Sherekea, Golden Purple, Desiree, Dutch Robjin, Arka, Nyayo, Roslin Tana, Roslin Eburu, Annet, Kerr's Pink and Kenya Baraka. Most of these varieties are adapted to the local environmental conditions and are used for domestic consumption and processing. Some of the varieties are, however, faced with low disease resistance hence the need for enhanced breeding to come up with high yielding cultivars that can withstand evolving diseases, pests and drought in the face of the ever changing climate.

Potatoes are grown by over 800,000 farmers on 160,000 hectares and are valued at US$ 0.54 billion per annum at the consumer level. The productivity is estimated to be 7 tons/hectare (MoA, 2007). Although production fluctuates, it has increased over the last decade from 1.2 to 2.4 million tons in 2003 and 2011, respectively (FAOSTAT, 2011). The Kenyan potato value chain indirectly employs 2.5 million people including researchers, extension agents, seed inspectors, transporters, market agents and processors among others. The crop is no doubt a major player in the country's economy ensuring livelihoods to many citizens in this regard. This paper looks at the entire value chain for both the seed and ware potatoes.

SEED AND WARE POTATO PRODUCTION

Potato production incorporates the seed and ware potato value chains. The ideal quality seed potato value chain consists of linked actors who produce seed potato of different generations starting from pathogen-free seed from public or private central research facility and with use of the seeds by ware producers (Wachira, 2013). Seed potato includes farm-saved seed (clean or positively selected, but not inspected by Kenya Plant Health Inspectorate Service, KEPHIS), basic seed and certified seed (which meets certification requirements). The Seeds and Plant Varieties Act (Chapter 326) sets the standards for basic and certified seeds. Inadequate seed supply is a major handicap to increased production as most farmers (96%) depend on their own saved seeds which generally lead to lower yields (Seed Potato Master Plan, 2010). Potatoes harvested are sold at harvest with storage for future sale rarely done despite the huge gluts and low prices at harvest followed by scarcity and high prices a few months later, just before the next planting season.

VALUE ADDITION: CURRENT PRACTICES, CHALLENGES AND OPPORTUNITIES

As a perishable crop, potato requires careful and proper post-harvest handling strategies in order to fulfill its role in the value chain. Care is to be taken during harvesting, transportation and storage. During harvest seasons, the crop is found in plenty and farmers are forced to sell the produce sometimes at a throw away price. This is perpetuated by lack of adequate capacity and cold storage chains that could prolong the shelf-life (Walingo et al., 2004). Post-harvest value addition is a viable means of not only assuring producers of higher return but also ensures availability of potato products on and off production seasons. Processing most certainly converts potatoes into less bulky and less perishable products that are easily transported.

The main products in the ware potato value chain are fresh potatoes, organically grown potatoes, pre-packed potatoes, pre-peeled, pre-cut and chilled, washed potatoes and baby potatoes (Wachira, 2013). Majority of the households in potato growing zones and a number of urban dwellers use the tubers as important accompaniment to diverse dishes. On the other hand, potato crisps and French fries (chips) are important processed products that form greater parts of menus in restaurants and hotels in major urban centers with crisps assuming large share in the supermarkets, kiosks and roadside shops (Walingo et al., 2004; Abong' et al., 2010a). Potato chips and crisps are the most industrially produced potato products in Kenya. There are over 20 crisps processing industries producing over 30 brands in Nairobi with majority (>80%) being small scale (Abong' et al., 2010a). This is a significant increase of the number of processors over the 5 recorded in 1979 by Durr and Lorenzl (1980). For many hotels and restaurants, French fries are prepared mainly from freshly harvested potatoes, while a few hotels and industries emulate the concept of frozen French fries that are mainly imported (Walingo et al., 2004). Other potato products traded in the Kenyan market include frozen chips, *chevda* and *bhajias*. On the other hand, processing of other products such as flakes, flour, dehydrated cubes and mashed potatoes remain feasible but untapped despite having great potential.

MARKETING OF POTATOES

Ware potatoes are transported in lorries to wholesale and retail markets of urban centers in Nairobi, Mombasa, Kisumu and Nakuru. Most grading is done on the market using fresh produce quality standards as a reference point. Despite seasonality of the crop, farmers of the fresh produce make substantial income. They are, however, faced with a number of challenges. Although the legal limit for ware potato bags is 110 kg, middle men use extended bags often over 150 kg to fleece farmers in times of oversupply leading to a loss of over US$ 10 million per annum to the growers (Kasina and Nderitu, 2013). Due to lack of storage facilities, farmers often sell at low prices at harvest while consumers pay high prices 2-3 months later due to low supply of potatoes before the next harvest. Good infrastructure is still needed in the growing seasons to aid in transportation of the fresh produce to marketing centres (NPCK, 2013).

Demand for processed potato products has been on steady rise over the years (Abong' et al., 2010b). Consumers, however, still lack adequate market information on the various products (Abong' and Kabira, 2012). Quality of varieties used for processing needs improvement to attract investors into value chain (Wachira, 2013). The ware potato value chain in Kenya is complex and inefficient (Kasina and Nderitu, 2013). The amount of money lost throughout the chain due to lack of transparency in the trade was estimated at about US$ 23 million in 2010.

CONCLUSION

Potato is now considered by the Ministry of Agriculture, Livestock and Fisheries as a possible alternative crop to maize which is having problems with incurable diseases from unknown sources. Many farmers in maize growing areas are now adopting potatoes for food and income security reasons. The Kenya National Potato Association (KENAPOFA) and the Kenya National Potato Council in collaboration with the Ministry of Agriculture, Livestock and Fisheries and related ministries are expected to move the potato value chain to competitive levels similar to other major cash crops such as tea and coffee.

REFERENCES

Abong' GO, Okoth MW, Imungi JK, Kabira JN (2010a). Characteristics of the potato crisps processing industry in Kenya. Journal of Animal Plant Science, 8(1): 936- 943.

Abong' GO, Okoth MW, Imungi JK, Kabira JN (2010b). Consumption patterns, diversity and characteristics of potato crisps in Nairobi, Kenya. Journal of Applied Biosciences, 32: 1942–1955.

Durr G, Lorenzl G (1980). Potato production and utilization in Kenya. International Potato Center (CIP).

FAOSTAT (2011). Food and agricultural commodities production. http://faostat.fao.org/site/339/default.aspx. Accessed on March 5, 2013.

Gildemacher PR, Kaguongo W, Ortiz O, Tesfaye A, Woldegiorgis G, Wagoire WW, Kakuhenzire R, Kinyae PM, Nyongesa M, Struik PC, Leeuwis C (2009). Improving potato production in Kenya, Uganda and Ethiopia: A System diagnosis. Potatao Research, 52: 173–205.

Kasina M, Nderitu JM (2010). Policy implementation and its economic impact on potato marketing value chain in Kenya. East African Agricultural and Forestry Journal, 76 (3-4): 161-168.

Ministry of Agriculture (2007). Challenges in potato research In: The National Policy on Potato Industry presentation during the potato stakeholders meeting at KARI headquarters, Nairobi, Kenya.

MoA (2009). Mainstreaming the potato crop from orphan crop status. Proceedings of Round Table Africa (RTA) potato stakeholders' workshop. Ministry of Agriculture, May 08, 2009, Nairobi, Kenya.

Nema PK, Ramayya N, Duncan E, Niranjan K (2008). Potato glycoalkaloids: formation and strategies for mitigation. Journal of Science of Food and Agriculture, 88:1869–1881.

NPCK. (2013). The value chain challenges. www.npck.org. Accessed on March 05, 2013.

Seed Potato Master Plan (2010). Seed potato subsector in Kenya, Ministry of Agriculture, Livestock and Fisheries, Nairobi, Kenya.

Wachira K. (2013). A roadmap for revitalization of the potato sub-sector in Kenya. Unpublished Kenya National Potato Council, KARI-NARL, Nairobi, Kenya.

Walingo A, Lung'aho C, N'gang'a N, Kinyae PM, Kabira JN (2004). Potato marketing, storage, processing and utilization in Kenya. Proceedings of 6th Triennial Congress of the African Potato Association, Agadir, Morocco 5-10 April 2004.

5. The potential of nixtamalization of maize to improve safety of consumption, nutritional value and processing property

Jasper K. Imungi[1*]
[1]Department of Food Science, Nutrition and Technology, University of Nairobi, Nairobi.
[*]*Corresponding author: imungijk@yahoo.com*

SUMMARY

Maize is the main staple for most of the communities in this Country. Contamination with mycotoxin has lingered as a public health concern each year, worse in some years because the toxicity turns fatal. In communities that consume maize solely, there have been reported cases of pellagra. Finally, the diversity of dishes consumed from maize is delimited by the poor functionality of the maize carbohydrates and proteins. This paper reviews the potential of nixtamalization or alkali/lime cooking in reducing the risk of mycotoxin poisoning, and improving the nutritional and processing properties of maize and maize products. The forms in which maize is prepared for consumption, including nixtamalization, are reviewed with a view to determining other dietaries that can be introduced to increase diversity. The major nutritional and health benefits that accrue from maize consumption are assessed with a view to improving them, and the change in the functionality of the maize protein and starch assessed with a view to determining the additional products that can be processed from maize and products. The paper also assesses the feasibility of production of nixtamalized maize products at the household and commercial or industrial levels. The paper concludes that nixtamalization of maize has great potential especially at the village level to help reduce risk of aflatoxin exposure as well as release niacin and make the product more nutritious. It is also possible to scale nixtamalization to commercial levels.

INTRODUCTION

Maize is the principal staple food for majority of Kenyans (Muriuki and Siboe, 1995; Shephard, 2008). It is consumed at the rate of 400 g per person per day (Kimanya et al., 2008) with consumption per capita of 98 kg (Nyoro et al., 2008). Maize contributes up to 36% of calorie provision to Kenyans (FAO, 2004).

Maize in Kenya is prepared and consumed in various forms. Ugali (a paste from maize meal and hot water) is the main dish. Other dishes include: 1) githeri (boiled mixture of maize kernels and beans in varying proportions where maize is always the major component). Githeri is occasionally stewed with either potatoes or green leafy vegetables; 2) *Muthokoi* (decorticated maize kernels) cooked either alone or in a mixture with beans or green leafy vegetables); 3) Irio (a cooked mashed mixture of maize and beans or pulses, green leafy vegetables potatoes and/or green bananas; and 4) uji (porridge mainly prepared from maize meal or flour. Uji can also be prepared by soaking the maize kernels in water, then grinding in an attrition mill or pounding in a mortar with a pestle. The nutrition of millions of Kenyans is therefore to a large extent dependent on maize.

Alkali cooking of maize (nixtamalization) is not very uncommon. The Kalenjin community in the Rift Valley use alkaline plant ashes for decortication of maize kernels (Kirui, 2011). The Kisii in western Kenya use alkaline herbal salts for cooking maize and beans in githeri. The maize kernels cooked this way change colour to light yellow and are said to be much tastier than those cooked with plain water (Onyango, 2012). Many communities in the country cook maize porridge and add alkaline ashes, alkaline salts or soda ash. Porridge cooked this way turns yellow and it is said to be tastier than porridge cooked without alkali.

BOX 1: NIXTAMALIZATION

Nixtamalization involves cooking and steeping of maize or other grain in an alkaline solution, such as calcium hydroxide, *lime, (*or ash or sodium carbonate). Nixtamalization softens the hull and partly detaches it from the rest of the kernel so that it can be rubbed off and washed away. The process helps to transform the kernels into cohesive dough. An advantage of the procedure is that it releases much of the bound niacin that we can absorb and benefit from.

Alkali cooking of maize has nutritional advantages. The niacin which occurs in bound form is released and can be better utilized by the body. Moreover, cooking also helps to reduce phytate content and improves the funtionality of the protein (Bressani et al., 1997). Moreover cooking of the kernels after decortication reduces levels of mycotoxins (Garcia and Heredia, 2006), a major problem in maize in the tropics (Strosnider et al., 2006; Wagacha and Muriithi, 2008).

NIXTAMALIZATION OF MAIZE IN KENYA

Alkali cooking of maize is not uncommon in Kenya. Decortication by cooking with alkali is practiced by the Kalenjin community. The maize kernels are mixed with alkaline ash from burning of certain woody materials known to produce the type of ash. Water is added and the maize soaked for a while, then cooked for about 45–60 minutes. Cooking loosens the skin which is removed easily by rubbing the kernels between the hands. The decorticated kernels are further washed till the wash water becomes clear from yellow. The kernels are then further cooked in water to remove the residual yellow colour. This product is still known as *muthokoi* and is similar to that obtained by pounding with mortar and pestle. The *muthokoi* is then either cooked right away or dried for later use (Kirui, 2011).

Other alkali cooking of maize in Kenya include: 1) cooking of githeri (maize and bean mixture) with either alkaline ash or alkaline herbal salt among the Kisii community of western Kenya (Onyango, 2012). The alkali or alkaline earth metals responsible for the alkalinity of the ash have not yet been scientifically determined. 2) Addition of alkali as either the alkaline ash or soda ash during cooking of maize meal or maize flour porridge. Cooking with the addition of the alkali changes the product colour to yellow and improves the taste of the food.

The alkaline ashes and salt are prepared commercially and sold in markets especially in Western Kenya by the Luo and Luhya communities. The ash is prepared by burning dry plant residues such as the banana pseudo-stem, maize cob, bean trash and many other residues from unidentified trees and shrubs known by the communities. Other uses of the salts and the ashes are in the cooking of over mature vegetables to reduce toughness, cooking of beans that have developed the hard to cook property and cooking of tough meat. The ashes and salts are also added during cooking of okra to bring out the sliminess.

Unconfirmed reports indicate that the preference for *muthokoi* prepared by pounding in mortar and pestle, commercial mills, and by nixtamalization decreases in the order: nixtamalized → pounding → machine. In comparing the three methods of decortication also, it can be seen that the mortar and pestle is a more labour intensive and leads to undesired breakage of kernels, while machine decortication is more capital intensive than nixtamalization.

NIXTAMALIZATION TO IMPROVE MAIZE NUTRITION

In Kenya, improvement of maize nutrition can directly lead to improvement of community nutrition. There are some communities in the country that will subsist on maize solely without possible supplementation with legumes or green leafy vegetables, especially during food scarcity. Such communities include the Kamba of Eastern Kenya. Cases of pellagra have been reported in this area (Daily Nation, 2009; Imungi, 2009). Niacin whose deficiency in diets causes pellagra is found in maize bound as niacinogen, and is therefore not available to the body. Moreover the other source of niacin in maize, the amino acid tryptophan is severely limiting. Cooking of maize with alkali helps to release niacin from niacinogen so that it becomes available to the body. Cooking also helps to reduce the level of phytates and improves the quality of the protein.

DAILY NATION
Tuesday August 25, 2009

PELLAGRA DIET

JASPER K. IMUNGI

Why Kenyans are ailing from eating maize alone

THE AUGUST 12 EDITION OF The *Nation* ran a story that communities of Ndalani in Makindu District were suffering from pellagra-type ailment because of depending on meals of maize alone.

I say pellagra-type because, pellagra, the disease associated with deficiency of the so-called Vitamin B3, is sometimes complicated by parallel occurrence of deficiency symptoms of other nutrients.

The chemical names for this anti-pellagra factor or Vitamin B3 are nicotinic acid, nicotinamide and niacinamide, forms in which it is available from food.

For the lay persons, it is known by the term niacin. Niacin can also be synthesised in the body from the amino acid tryptophan if the amino acid is present in the meal consumed.

Usually, 60mg of tryptophan will be converted to 1mg of niacin, a very wasteful way of utilising the amino acid. Tryptophan is present in beans and other legumes and complements very well with tryptophan, a meal containing maize and beans like *githeri*.

THE RECOMMENDED DAILY ALLOWance (RDA), or the amount of niacin to be taken through food by humans, varies from 5mg for infants to 20mg for adult males.

Pellagra is a very common nutritional deficiency disease all over the world, especially in underdeveloped countries among communities that depend on predominantly maize-based dietaries low in niacin and quality protein.

The protein of maize (zein) is virtually devoid of tryptophan. Recently, however, the International Centre for Improvement of Maize and Wheat based in Mexico has embarked on biotechnological development of quality protein maize, which is heightened levels of tryptophan and lysine, two amino acids lacking in cereals.

Beans or vegetables add a great deal of nutritional value to a plate of maize.

When the production of this maize takes ground properly, its consumption will not result with the problems associated with pellagra. The main sources of niacin in diets include lean meats, liver, poultry, milk, tomatoes, and several green leafy vegetables, including the traditional vegetables commonly consumed by our communities.

It was observed some time ago that while those communities subsisting on maize diets alone were suffering pellagra, the Mexicans were not. This was traced to the way they were preparing maize for consumption as tortillas, a kind of thin pancake.

They were incorporating alkali in the formulation of the batter. This led to the discovery that niacin occurs in the maize in bound form as niacinogen, from which it cannot be dislodged by the digestive system. However, on heating in the presence of alkali, the niacin was released.

To benefit from the maize niacin, therefore, it would be advised that the food be cooked with addition of small quantities of magadi soda or such other alkali.

One such alkaline solution is the watery extract of ash from dried bean trash, banana peels and leaves and banana pseudo-stem, though not exclusively, used by the Luhya and Luo communities of western Kenya (*Omushelekha/kat kulo)* for cooking such vegetables like bush okra (*omurere/mrenda/apoth*).

Of course, the release of niacin from the niacinogen will be more effective the more the maize is divided.

Preparation of food like ugali or uji, especially the latter, will release niacin better than cooking of whole maize kernels.

The other advice would be to cook the maize diet (whole kernels and decorticated kernels or *muthokoi*) with green leafy vegetables to benefit from the niacin from the vegetables.

POUNDING MAIZE FOR *MUTHOKOI* with green leafy vegetables or any leaves that are not known to be toxic (not necessarily used as vegetables) will also achieve the same purpose.

When alkali is used in the cooking, the old and brisk vegetables which are considered difficult to cook become softened by the action of the alkali.

I am aware that during the drought period, even the vegetables are very scarce. However, it is also during such times that we remember the diversity of plants that African communities have utilised as vegetables and where they are likely to grow during the periods.

Prof Imungi teaches food technology in the department of Food Science, Nutrition and Technology, University of Nairobi.

Decortication of maize with alkali is not common among the Kambas, although the term *muthokoi* used for decorticated maize has emanated from their language. The Kambas prepare *muthokoi* by pounding the maize kernels mixed with water to soften the skin in a mortar and pestle. Occasionally, maize that is considered too brittle is mixed with green leafy vegetables in addition to water during pounding.

It has been observed that many school feeding programs especially in the poor rural areas utilizes maize products. Githeri containing very little beans is a very common diet. These children live in the borderline between pellagra (Box 2) and niacin deficiency. Even when ugali is the meal served, it is often with very little vegetable based accompaniment.

BOX 2: PELLAGRA

Pellagra is a vitamin deficiency disease caused by niacin (vitamin B_3) deficiency resulting from inadequate dietary intake of nicotinic acid and/or its precursors, and the essential amino acid tryptophan. This defieciency affects the skin, digestion and nervous system and is clinically manifested by photosensitive dermatitis, diarrhoea and dementia.

NIXTAMALIZATION AS A MEANS TO REDUCE AFLATOXIN LEVELS IN MAIZE

Acute aflatoxicoses associated with eating home grown maize which has been stored under damp conditions has been a perennial problem in Kenya since the late 1970s, especially in the eastern and central parts of the country (Ngindu et al, 1982; CDC, 2004; Lewis et al., 2005). It is believed that thousands of people in these areas are each year exposed to aflatoxin at sublethal doses, putting them at risk of chronic toxicity (Fandoha et al., 2003; Lewis et al., 2005; Wagacha and Muthomi, 2008). It is in this area that the communities are accustomed to eating *muthokoi* made by pounding in mortar and pestle.

The outbreak of aflatoxin poisoning in 2004 in Kenya remains the most severe recorded worldwide so far. The outbreak covered more than seven districts and resulted in 317 hospitalized and 125 reported deaths. Most of the populations were obviously exposed to sub-lethal doses of the toxin. The maize associated with this outbreak was harvested in February during off-season early rains. The levels of aflatoxin B_1 in the maize was found to be 4,400 ppb, over 400 times more than the tolerance for the country (Baumgartner et al., 2005; CDC, 2004). Other outbreaks were reported in 2005 and 2006 and resulted in 30 and 7 deaths respectively.

In 2010 more than 23 million bags (1 bag = 90 kg) of maize grown by smallholder farmers was condemned and bought by the Government at 30% of the prevailing price for destruction by incineration. The maize was contaminated with aflatoxin at levels above the tolerable limit. Of course the maize that stayed back in the farmers' homes for domestic consumption contained similar levels. There are no recommended simple methods of decontamination for these people. Aflatoxin contamination is a topical phenomenon. Decortication by whatever method therefore reduces the level of contamination considerably. Kirui and Imungi (2011) found that cooking of maize in alkaline infusions from maize cob ash reduced the levels of aflatoxin B_1 initial average level of 86 ppb to level of 6.1 ppb in *muthokoi*.

CONCLUSION

Nixtamalization of maize in Kenya can have ample nutritional and health benefits. It can help solve the problem of niacin deficiency now prevalent among maize eating communities and the incicdences of aflatoxin poisoning, which is almost perennial in the eastern and central regions of the country. Nixtamalization or cooking of maize and maize products with alkali in Kenya is not unknown, and the tastes of the products from that kind of cooking are acceptable by some Kenyan communities. Scaling up the practice will therefore not be a major challenge.

REFERENCES

Azzi-Baumgartner E, Lindblade K, Gieseker K, Schurz Rogers H, Kieszak S, Njapau H, Schleicher R, McCoy, LF, Misore A, DeCock K, Robin C, Slutsker L. (2005). Case - Control study of an acute aflatoxicosis outbreak, Kenya, 2004. Environmental Health Perspectives, 113:1779 – 1783.

Bressani R, Rooney LW, Saldivar SOS. (1997). Fortification of corn masa flour with iron and/or other nutrients: A literature and industry experience review.

CAC. (2001). Joint FAO and WHO food standards program. Codex Alimentarius on food additives and contaminants. Codex Alimentarius Commission, 33rd Session, Netherlands.

Daily Nation. (2009). Kenyans are ailing for eating maize alone. Daily Nation 2 August 2009.

Fandohan P, Hell K, Marasas WFO, Wingfield MJ. (2003). Infection of Maize by Fusarium species and contamination with fumonisin in Africa. African Journal of Biotechnology, 2:570–579.

FAO (2006). Food security situations. Food Consumption and population growth. Food and Agricultural Organization, Rome, Italy

Garcia S, Heredia N. (2006). Mycotoxins in Mexico: Epidemiology, management and control strategies. Mycopathologia, 162:255 - 264.

Imungi JK. (2009). Why Kenyans are ailing for eating maize only. Daily Nation Kenya, August 25, 2009.

Kimanya ME, De Meulenaer B, Tiisekwa B, Ndomondo-Sigonda M, Devlieghere F, Van Camp J, Kolsteren P. (2008). Co-occurence of fumonisins with aflatoxins in home-stored maize for human consumption in rural villages in Tanzania. Food Additives and Contaminants, 25: 1353 – 1364.

Kirui CC. (2011). Personal communication. National Cereals and Produce Board, Nairobi, Kenya.

Kirui CC, Imungi JK. (2011). Use of local alkaline ash in detoxification of aflatoxin in maize. Unpublished data. Department of Food Science, Nutrition and Technology, University of Nairobi, Kenya.

Lewis L, Onsongo M, Njapau H, Schurz Rogers H, Luber G, Kieszak S, Nyamongo J, Backer L, Dahiye AM, Misore A, DeCock K, Robin C (2005). Aflatoxins contamination of commercial maize products during an outbreak of acute aflatoxicosis in Eastern and Central Kenya. Environmental Health Perspectives, 113:1763 – 1767.

Muriuki GK, Siboe GM. (1995). Maize flour contaminated with toxigenic fungi and mycotoxins in Kenya. African Journal of Health Sciences, 2:236 – 241.

Ngindu A, Kenya PR, Ocheng DM, Omondi TN, Ngare W, Gatei D, Johnson B, Ngira JA, Nandwa H, Jansen AJ, Kaviti JN, Siangok TA. (1982). Outbreak of acute hepatitis by aflatoxins poisoning in Kenya. The Lancet, 319:1346 – 1348.

Nyoro JK, Kirimi L, Jayne TS. (2004). Competitiveness of Kenyan and Ugandan maize production: Challenges for the future. Tegemeo Insitute, Kenya.

Onyango CM. (2012). Personal Communication. Department of Crop Science and Protection. University of Nairobi, Kenya.

Shephard GS. (2008). Risk assessment of aflatoxins in food in Africa. Food Additives and Contaminants, 25(10):1246 – 1256.

Strosnider H, Azziz-Baumgartner E, Banziger M, Bhat RV, Breiman R, Brune MN, DeCock K, Dilley A, Groopman J, Hell K, Henry SH, Jeffers D, Jolly C, Jolly P, Kibata GN, Lewis L, Liu X, Luber G, McCoy L, Mensah P, Miraglia M, Misore A, Njapau H, Ong CN, Onsongo MTK, Page SW, Park D, Patel M, Phillips T, Pineiro M, Pronczuk J, Rogers HS, Rubin C, Sabino M, Schaafsma A, Shephard G, Stroka J, Wild C, Williams JT, Wilson D. (2006).

Workgroup Report: Public health strategies for reducing aflatoxin exposure in developing countries. Environmental Health Perspectives, 114(12):1898 - 1903.
Wagacha JM, Muthomi JW. (2008). Mycotoxins problem in Africa: Current status, implications to food safety and health and possible management strategies. International Journal of food Microbiology, 124:1 – 12.

6. The case for food security imperatives in natural resource development and management in Kenya - constitutional, policy and legislative perspectives

Nathans Browne JP*
International Policy Research Centre, Nairobi, Kenya.
**Corresponding author: browne@internationalpolicyresearchcentre.org, browne@iprec.org.*

The Constitution of Kenya [2010] makes provision for 'adequate food' as a right to which every citizen is entitled, declaring 'freedom from hunger' as a corollary thereto, pursuant to Article 43 thereof. This paper seeks to make the case for policy and legislative interventions towards effective implementation of this particular provision of the Constitution. Particular emphasis is placed on specific policy and legislative instruments that govern the realm of natural resources, with the paper advancing the argument in favour of harnessing and fully utilizing both the high potential agricultural land as well as the arid and semi-arid lands (Box 1) in a prudent and sustainable manner towards achieving a reasonably stable level of food security in Kenya.

BOX 1: KENYA'S LAND MASS

Kenya's 582,647 km^2 territorial land mass is inhabited by about 43 million people. This land mass can be subdivided into 20% high potential agricultural land and 80% constituting the arid and semi-arid lands Government of Kenya statistics indicate that the 20% portion of the country's territory caters for nearly 80% of the population.

Food production is, primarily, a consequence of agricultural practices which lead towards a sustainable economy for any nation. Kenya places special emphasis on this sector as a key pillar upon which the economy has remained predicated since pre-independence days. The sector encapsulates crop production, livestock development as well as nurturing the ever-important fisheries sector. The Agriculture Act, Cap 318 Laws of Kenya, remains the key legislative bed-rock in this regard.

The paper examines other policy and legislative imperatives which include the Environmental Management and Coordination Act [EMCA] 1999 as the over-arching instrument. Further consideration is given to provisions governing water, forests, mining, oil and natural gas, wildlife, land, livestock, climate change, among several other laws and policies.

The paper makes the case that use of drought-resistant crops, appropriate seed varieties as well as proper land use planning must be accorded the attention they deserve. Adverse effects on the environment as a result of climate change are given due consideration, and the argument is made that communities must be introduced to adaptation methods for the purposes of achieving food sufficiency, while the state is obligated, under international legal instruments, now a part of national laws, to employ and use requisite mitigation measures in this regard.

The paper seeks to address the subject-matter within the Constitutional context of Kenya's new devolved system of governance, under which the national government is assigned national economic policy and planning, the coordination of planning by the Counties, protection of the environment and natural resources including fishing, wildlife, water protection, energy and agricultural policy responsibilities. County governments have been assigned specific responsibilities which include crop and animal husbandry, implementation of specific national government policies on natural resources and environmental conservation, soil and water conservation as well as forestry.

Overall, the paper makes the conclusion that proper use, development and management of Kenya's natural resource base would, by and large, become a major contributing factor towards adequate food security for the country's nearly 43 million people, thus enabling the citizens to produce enough for both domestic use as well as retaining the capacity to export part of what constitutes surplus quantities.

The use and application of proper seed and crop varieties must, of necessity, be given priority consideration within the framework of production, processing and consumption in Kenya.

Without unnecessary regard to technicality and philosophical attributes, food security should be defined to include the sum total process of prudent development and management of economies, whether local, national, regional or global, with the net effect of self-sufficiency with regard to the stability and cohesive existence of populations, through sustainable exploitation, conservation and management of factors of production and scale-by-scale practices of consumption.

In a sense, therefore, food security entails the ability on the part of a country to provide for its citizenry, in terms of sufficiency of all that appertains to feeding and nutritional process in the food matrix and supply chain. It is often said that a nation that is not self-suuficient in food supplies risks the vagaries of conflict, unrest, despondency and, as a consequence, civil disobedience. As a corollary, the population of a country that is not self-sustaining in terms of food production, consumption and supply stands the potential to suffer hunger, disease and malnutrition, among several other maladies.

The realm of natural resources spreads far, deep and wide; the various components that constitute this realm include marine resources, those in the wild - wildlife and forests - water, minerals, oil and natural gas, plant and plant species. These components are, generally speaking, well articulated within the body-fabric of the 1992 Convention on Biological Diversity, details whereof are highlighted in succeeding paragraphs.

The main argument postulated in this paper is the pre-supposition that whereas a country will feed its people through routine processes of planting and harvesting staple foods and related items and other supplements, the case must be firmed up for the need to harness, develop and sustainably manage existing categories of natural resource components as additional as well as alternative sources of sustenance and subsistence. How this is to be done remains the sole prerogative of each individual country, save only that policy and legislative imperatives must be considered in order to have the assurance for a secure and reliable system of production, supply, consumption and, where appropriate, for export purposes.

By way of introductory remarks on this subject, key attributes, therefore, that would come in handy to safeguard the overriding interests of the citizenry include proper agricultural practices, fully harnessing existing or untapped agricultural potential which, if well and properly made use of, would enhance and improve food security, with the concomitant effect of wealth creation and job provision for the general populace.

A country, any country for that matter, is well advised to initiate major irrigation mechanisms particularly within major low productivity areas, such as the arid and semi-arid areas; it is possible to cultivate horticultural food components to meet demand first for local consumption and, where in excess, for export purposes.

Support for the farming community in any country would range from the general agriculture-based practices, fisheries, dairy as well as sugar belt development and management. The foregoing, essentially, can best be addressed through use of modernized infrastructure, development of networks through which the farmers in the rural areas can easily deliver their merchandise to the market such as via road and rail connections. Provision of 'cooling' facilities can only be best achieved through provision of proper, reliable and affordable electrification mechanisms, whether through wind, geothermal, hydro or electric power.

This paper seeks to examine Kenya's relatively rich and abundant natural resource base, putting forward the argument that the key pillars upon which the country has set herself for advancement and growth must be strengthened through constitutional, policy and legal architectural formulation, with a view particularly to achieving the goals and objectives spelt out in the country's development blue-print of Vision 2030.

The case for Kenya's food security cannot be appropriately made out without an opportunity to review the country's long journey on the path to harnessing and consolidating the historical gains in the realm of her natural resource base. Pre-colonial policy and legislative architecture was largely an import within the framework of what was to be known as 'the 1897 inception clause', by which it was a rule of thumb that any (policies and / or) laws transported into Kenya were acceptable so long as the same did not offend, or were not repugnant to, morality and justice.

Arguably, it was not until 1963 when Kenya as a Republic took charge of a multi-dimensional policy and legislative process, spearheaded by the country's pre-independence-trained and well-educated local manpower, although by and large the process was an all-inclusive and participatory one, bringing on board all shades, opinions and views in all fields, including remnants of pre-colonial Kenya. The catapultic thrust of this framework was, essentially, to lay a firm basis for post-independent national planning development as well as national planning for years to come. It can therefore be safely stated that Kenya's national food security fulcrum was, by and large, crafted, developed and subsequently codified through consideration for economic imperatives, as the foundation for national development.

Events on the international stage further fuelled the fire for need to strike a balance between the protection, conservation and sustainable management of the environment on the one hand, and (national planning and) economic development on the other.

Nations of the world converged in Stockholm in June 1972, in what was to be known as 'the Stockholm Conference on Environment and Development', as a result of which all governments became aware of the limitations of the bio-sphere, hence the need to base economic development within the framework of the finite nature of the earth's finite natural resource endowment. The net effect of the 'Stockholm declaration' was, therefore, that all states must take urgent measures to ensure that consumption by the present generations must be based on sustainable utilization of natural resources, in order to, essentially, cater for generations yet to be born. It may be argued, and probably quite correctly, that the spirit behind the 1972 Conference was founded upon the effects of several catastrophes, including the 1970s oil crisis as well as waves of famine and drought across the African continent in general, and the horn of Africa in particular.

The philosophical foundation upon which the World Commission on Environment and Development, as crafted in 1987 under the 'Our Common Future' banner, confirmed the inter-dependence of both development and environmental goals. Matters were taken a notch higher when, again, nations of the world converged in Rio de Janeiro, Brazil, to articulate these emerging concerns for mankind, leading to the adoption of Agenda 21, as a platform for attaining sustainable development. Consequently, some critical international environmental instruments were generated, in particular the 1992 Convention on Biological Diversity' (CBD), and the 1992 United Nations Framework Convention on Climate Change (UNFCCC). In addition, African states, guided by their governments, facilitated and agitated for the negotiation and subsequent adoption of 'the United Nations Convention to Combat Desertification' [UNCCD].

As a core-relation to the events and activities at the global arena, Kenya's long journey to a credible constitutional, policy and legislative architecture in natural resource development and management can be traced to the country's long-cherished process of planning for development, commonly referred to as the 'development plan' mechanisms.

Foremost of the relevant measures undertaken was Sessional Paper No. 10 of 1965 on African Socialism and its application to planning in Kenya, whence the country gave recognition to the role of Natural Resources in Economic Development.

Sessional Paper No. 3 of 1976 set out the National Policy on Wildlife Management. Hearkening to the call arising out of the Stockholm Conference, Kenya injected the overriding philosophy into her 1974-1978 Development Plan, with its key bed-rock position being the buffer between human welfare and environmental conditions.

The 1979-1983 Development Plan was crafted with provisions containing perhaps a much more elaborate approach to environmental issues, giving natural resource conservation, protection, management, exploitation and use precedence over development activities.

The 1989-1993 Development Plan placed greater emphasis on serious concerns that degradation of natural resources would, as a consequence, affect economic development, growth as well as national planning for the country.

The 1994-1999 Development Plan was perhaps a lot more robust in its pronouncements. The Policy called for appropriate integration of environmental management considerations in all facets of policy-making, as well as legislative formulation, drafting and review.

It is noteworthy that the pivot upon which the 1994-1999 Development Plan was founded was the call for major legislative and institutional reforms in Kenya.

To reinforce the foregoing, the1995-1996 National Environment Action Plan [NEAP] process set out to address nearly all natural resource-related components; the process specifically made provision for a raft of key proposals and measures intended for effective environmental management, protection and conservation.

It will be remembered that through the NEAP process, a call was made for a review of various policies and laws in respect of Kenya's natural resource base, with a view to working towards the enactment of a major national legislation.

The NEAP placed emphasis on the need for Kenya to embed environmental considerations in national development, with a clear recommendation for a clear policy, legislative and institutional positioning towards effective management of the environment and national development.

From a constitutional perspective, Kenya, in response to the foregoing, enacted a new Constitution, the Constitution of Kenya 2010, in which matters of the environment, land and natural resources were given particular prominence and considerable supremacy, by reason of the Constitution itself being the supreme law of the land. Specific provisions within the framework of the Constitution are captured under Chapter 5 therein.

Article 69 [1] (a) of the Constitution states thus:

"The State shall ensure sustainable exploitation, utilization, management and conservation of the environment and natural resources, and ensure the equitable sharing of the accruing benefits".

It can be safely stated, accordingly, that the constitution of Kenya 2010 makes clear provision on natural resource development and management with a clear intent on affirming the usefulness of the country's natural resource base for the purposes of assuring food security for her citizens.

From the foregoing, the following can be stated:

> By imposing an obligation on the State, the Constitution sets out to affirm the essentials of a commitment on the part of the state to its citizens, by reason of which it becomes incumbent upon the administrative and governance structures of the state to enforce this provision without fail
>
> The constitution makes provision for all aspects in relation to the natural resource base to be conducted in a sustainable fashion; sustainability here means the need to ensure that while use and exercise of the right to use natural resources is necessary for present generations, there is need for sufficient percentages to be spared for the use by generations yet to come.

The constitution places a major obligation on the part of the state to establish mechanisms through which any and all benefits [ideally entitlements] are shared and distributed equitably.

Without question, therefore, the state has a major obligation to ensure that all three [above] and several other imperatives are carried through, as a consequence of which matters of food security can safely be said to be within clear national focus and grasp, from the highest possible level of supremacy of legal demand.

To buttress the case for the economic interest of the individual, the Constitution makes the following provision, under Article 46 [1] (c):

"Consumers have the right to the protection of their health, safety and economic interests".

With particular regard to basic aspects of the right to food, the constitution makes the following provision, under Article 43 [1] (c):

"Every person has the right to be free from hunger, and to have adequate food of acceptable quality"

Aware that the country has since (the inception clause and) independence negotiated, acceded to, signed and/or ratified several international instruments such as treaties, conventions and agreements, the framers of the constitution of Kenya 2010 took cognizance of the importance and significance of international obligations.

Accordingly, Article 2 [5] states as follows:

"The general rules of international law shall form part of the law of Kenya".

This provision is reinforced further through the provisions of Article 2 [6], which states as follows:

"Any treaty or convention ratified by Kenya shall form part of the law of Kenya under the Constitution".

As has already been observed, the foregoing provisions of Article 2 [5] and 2 [6] of the Constitution of Kenya do invite for due and appropriate consideration of the letter and spirit of, *inter alia,* the following international instruments:

1. The Convention on Biological Diversity (CBD)
2. The United Nations Framework Convention on Climate Change (UNFCCC)
3. The United Nations Convention to Combat Desertification (UNCCD)

Bio-diversity is the main body-constituency of natural resources and its related components. Vagaries of climate change are associated with the hazards that befall farmers who find it difficult to produce under harsh changing weather patterns. Under the 2005 Kyoto Protocol to the Convention on Climate Change, measures with regard to mitigation and adaptation are codified, a clear indication, therefore, that the UNFCCC is a key international instrument in relation to food security. Under the Protocol, matters of carbon sequestration and carbon trading are highlighted. With regard to desertification, it is noteworthy that good seed varieties are a major bed-rock for any meaningful farming and planting season, affording the farmers of Kenya high yields for local and national consumption and supply, towards national food security.

The provisions of the CBD in particular were domesticated into the Environmental Management and Co-ordination Act [EMCA] 1999, with key highlights including:

1. General measures for conservation and sustainable use
2. Identification and monitoring of bio-diversity components
3. *In-situ* and *ex-situ* conservation
4. Sustainable use of biological diversity components
5. Research and training

Following the enactment of the Constitution of Kenya 2010, the country has since embarked on major policy and legislative reforms, as envisaged under the Sessional Paper pronouncements as already explained. Foremost in all these has been the 'collapsing' of close to 100 pieces of legislation that have for long governed the land regime in Kenya. Consequently, there has been enacted the following key laws:

1. The Land Act 2012
2. The Land Registration Act 2012
3. The National Land Commission Act 2012.

The net effect of these laws is that natural resource management and development has been given a clearer sense of legal direction. The law has now clarified what is private land, public land and community land. In addition to these three, the National Land Policy - Sessional Paper No. 3 of 2009 constitutes one of the most progressive policy platform upon which a framework on land management rests. These policy and legislative imperatives provide a well-synchronized mechanism through which key natural resource components may be harnessed, developed and managed. The land regime legislative and policy considerations lend great support to initiatives that are on-going within the purview of several other pieces of legislation that are currently under review, in order to bring the same in consonance with the provisions of the Constitution of Kenya 2010, and they include the following:

The Mining Act, Cap 306 (1st October 1940)

The Wildlife (Conservation and Management) Act, Cap 376 (13th February 1976)

The Agriculture Act, Cap 318 (1st July 1955, or, for the former Northern Province, 18th June, 1963)

The Petroleum Act, Cap 116 (31st August 1948)

In conclusion, it has to be re-stated that the case for food security in Kenya must not be viewed strictly from the prism of crop science and its related multi-dimensional components. Natural resources have a greater role to play in complimenting good agricultural practices, such as modernization of the mechanics and technologies that under-pine the scope of operationalization and implementation thereof. Kenya is continuously discovering new deposits of iron ore, gold, copper, manganese, coal, oil and natural gas, among several others. While it is important to place emphasis on good animal husbandry and agriculture-related practices, exploitation, use and utilization of natural resources, and the equitable sharing of accruing benefits can only but facilitate and compliment the country's net earnings towards food security for both present and future generations. And this is in consonance with the letter and spirit of the country's policy, legislative and constitutional pronouncements, as enunciated since the 1800s.

7. Enhancing β-carotene, ascorbic acid and sensory properties of potato crisps using carrot powder as a flavouring agent

George O. Abong'[1*], Jackson N. Kabira[2] and Michael W. Okoth[1]
[1]Department of Food Science, Nutrition and Technology, University of Nairobi, Nairobi, Kenya.
[2]National Potato Research Centre (KARI), Tigoni, Kenya.
[*]*Corresponding author E-mail: georkoyo@yahoo.com*

SUMMARY

Potato crisps are popular snacks consumed worldwide. Manufacture of crisps in Kenya is dominated by small-scale processors and there exist, depending on the market niche, a variety of flavours such as salted, tomato, chilli or cheese-flavoured. This study was designed to investigate the influence of carrot powder, as a flavouring agent, on β-carotene, ascorbic acid and sensory properties of potato crisps from three Kenyan potato varieties. Potato crisps from each variety were flavoured by applying carrot powder at the rates of 0%, 2.5% and 5% of crisps weight. Total carotenoids varied in raw tubers from 115.38 to 190.67 µg/100 g dry weight in Tigoni and Dutch Robjin, respectively. Total carotenoid content significantly ($P \leq 0.05$) increased on addition of carrot powder to crisps reaching highest levels of 921.42, 1,220.49 and 1,269.01 µg/100 g dry-weight, respectively. The levels of β-carotene were noticeably low in raw tubers and fried crisps, being undetected in Tigoni while it varied in Dutch Robjin (0.06-0.21 µg/100 g) to Cangi (0.11-0.26 µg/100 g). β-carotene and reduced ascorbic acid significantly ($P \leq 0.05$) increased with addition of carrot powder in all the varieties. Flavouring using carrot powder significantly ($P \leq 0.05$) reduced sensory scores on flavour, colour, oiliness and overall acceptability. Crisps flavoured at the rate of 0.25% were acceptable irrespective of the variety. On the other hand, all crisps flavoured at 0.5% were unacceptable to panelists. Carrot powder can therefore be used as a flavouring agent to provide a natural source of vitamin A to consumers.

INTRODUCTION

Potato crisps are popular snacks consumed all over the world. In Kenya, consumption of crisps has significantly increased in the past decade, especially in the urban centers (Abong' et al., 2010a). The local manufacture of crisps is dominated by small scale processors who use a variety of flavours such as salt, tomato, chilli or cheese-flavoured among others depending on the market niche (Abong' et al., 2010b). Apart from added flavours, the characteristic flavour of any particular crisps is influenced by the potato cultivar, frying oil and temperature (Martin and Ames, 2001). Flavour precursors such as sugars, amino acids and lipids are present in raw potatoes. Sugars and amino acids produce maillard reaction compounds while lipids are degraded during processing at elevated temperature to achieve distinct flavours (Jansky, 2010).

Carotenoids, including β-carotene which is a vitamin A precursor, are useful as antioxidants that protect the body against free radicals and hence help prevent diseases such as cancer (Edem, 2009). Depending on cultivar, the major carotenoids available in potatoes include lutein, zeaxanthin, violaxanthin and neoxanthin while β-carotene is normally found in small amounts (Dellapenna and Pogson, 2006; Burgos et al., 2009; Ezekiel et al., 2011). Deficiency of vitamin A intake is attributed to insufficient intake of the vitamin or its precursors and hence the need to increase intake.

Due to high consumption of potatoes in the world, many approaches have been used to increase vitamin A intake, potato and its products being one of the reliable vehicles. Andreu-Sevilla et al. (2009) demonstrated that using red palm oil as frying aid could increase the levels of β-carotene intake and hence vitamin A consumption (Johnston, 2009; Pedreschi et al., 2011). Through overexpression of bacterial phytoene synthase, carotenoids and β-carotene contents were increased six and eleven fold respectively in potato variety Desiree (Ducreux et al., 2005).

Natural flavours with added benefits of vitamin provision can be derived from available produce. Carrots are known to have very high levels of β-carotene which indicate high provitamin A activity in humans (Zude et al., 2007; Hayashi et al., 2012). Carrots provide the richest source of β-carotene for a number of people in the developed countries where its use is extensive (Berger et al., 2008). Production of carrot powder and its use as a flavouring agent in the highly consumed potato crisps does not only serve as a means of preserving carrots from fresh weight losses which they suffer quite often, but also increases vitamin A and reduced ascorbic acid intake for a large number of consumers in the market. This study was therefore designed to investigate the influence of carrot powder, as a flavouring agent, on β-carotene, ascorbic acid and sensory properties of potato crisps from three Kenyan varieties.

MATERIALS AND METHODS

Potato for processing

Two established potato varieties known for their good processing qualities (Tigoni and Dutch Robjin) were grown at the National Potato Research Center, Tigoni, Kenya in 2011 and 2012 under standard cultural conditions (Lung'aho and Kabira, 1999). After maturity, the crop was dehaulmed two weeks before harvesting and allowed to cure in a dark store under ambient air conditions (17-22°C, 84-94% relative humididty) for two weeks at the National Potato Research Center in Tigoni. The two varieties and an additional highly marketed variety (Cangi) were then processed into crisps in Tigoni.

Production of carrot powder

Fresh carrots were purchased from a local market in Nairobi. They were washed and cut into small pieces approximately 2-3 mm slices. The slices were water blanched at 95°C for 5 minutes before being dried in forced air oven at 85°C for 72 hours. Dried carrot cubes were ground using hammer mill (Condux D6450 Hanau, Germany) to pass through a sieve of 250 μm.

Production of flavoured crisps

Potato tubers were peeled and sliced using an automatic electric slicer (Hitech Systems, Saudi Arabia) to uniform thickness of 1.5 mm. The slices were thereafter washed in cold water to remove surface starch, dried with cloth towel and fried in an institution size, batch type, deep oil fryer (E6 ARO S.A., La Neuveville, Switzerland) in about 7 litres of "Chef" corn oil (Premier Oil Mills Ltd., Nairobi, Kenya) at 170°C until bubbling ceased. The fried slices were removed and excess oil drained off for 1 minute, cooled and packaged in polyethylene bags ready for flavouring. About 100 g crisps from each variety were flavoured by applying carrot powder at the rates of 0%, 2.5% and 5% of crisps weight. Effects of flavouring on ascorbic acid, β-carotene, oil content, colour and texture and sensory properties of flavoured crisps were analyzed.

ANALYTICAL METHODS

Total carotenoids and β-carotene

Total carotenoids and β-carotene were determined by modified Method No. 44 of the International Federation of Fruit Juice Products (IFFJP, 1972).

Reduced ascorbic acid

Ascorbic acid was determined by titration with 2, 6-dichlorophenolindophenol dye as described by Method No. 43.057 of AOAC (1980).

Crisps oil content

Duplicate 5 g potato crisps finely ground samples were accurately weighed and placed into thimbles and extracted in Soxhlet apparatus using analytical grade petroleum ether (boiling point 40-60°C) for 16 hours as described by Lulai and Orr (1979).

Colour measurement

Crisps colour was measured using a colour spectrophotometer (NF 333, Nippon Denshoku, Japan) using the CIE Lab L*, a* and b* colour scale as described by Abong' et al. (2011b) where 'L*' value is the lightness parameter indicating degree of sample lightness varying from 0 = black to 100 = white. The 'a*' is the chromatic redness parameter whose value means red colour when positive (+) and green colour when negative (–). The 'b*' is yellowness chromatic parameter corresponding to yellow colour when positive (+) and blue colour when negative (–).

Sensory evaluation

Coded samples were presented in duplicates to 20 panelists, all familiar with potato crisps. Panel members scored for colour, flavour, texture, oiliness and overall acceptability on a 7-point hedonic rating scale varying from 1 (dislike very much) to 7 (like very much). A score of 4 was the lower limit of acceptability.

Data analysis

One-way analysis of variance and least significant difference test for the variables was conducted using the Statistical Analysis System (SAS Version 9). Correlation coefficients were used to determine relationship between colour and texture determined in laboratory and sensory scores and differences at $P \leq 0.05$ were considered significant.

RESULTS AND DISCUSSION

Effect of carrot powder on total carotenoid, β-carotene and reduced ascorbic acid

The effect of carrot powder on total carotenoids, β-carotene and reduced ascorbic acid in potato crisps is shown in Table 1. Potato variety significantly ($P \leq 0.05$) influenced total carotenoids, β-carotene and reduced ascorbic acid. Potato variety interaction with carrot powder significantly ($P \leq 0.05$) influenced these parameters. Total carotenoids varied in raw tubers from 115.38 to 190.67 µg/100 g dry-weight in Tigoni and Dutch Robjin, respectively. Total carotenoids significantly ($P \leq 0.05$) increased on addition of carrot powder to crisps reaching highest levels of 921.42, 1,220.49 and 1,269.01 µg/100 g dry weight at 0%, 2.5% and 5% carrot powder, respectively.

The levels of β-carotene were noticeably low in raw tubers and fried crisps, being undetected in Tigoni while it varied in Dutch Robjin (0.06-0.21 µg/100g) to Cangi (0.11-0.26 µg/100 g). β-carotene, however, significantly ($P \leq 0.05$) increased with addition of carrot powder in all the varieties reaching highest levels of 274.88, 358.79 and 381.89 µg/100 g in Tigoni, Dutch Robjin and Cangi, respectively.

The values of β-carotene in raw tubers found in the current study compare with those reported by Brown (2008) who noted levels of 50-350 µg/100 g and 800-2,000 µg/100 g fresh-weight in white and yellow cultivars, respectively. On the other hand, β-carotene levels agreed with those reported by Bonierbale et al. (2009) and Burgos et al. (2011) who noted a variation in values from trace to 18 and 0-150 µg/100 g fresh-weight, repectively. Like other nutrients, carotenoid contents of potatoes depend mainly on the variety (Morris et al., 2004; Toledo and Burlingame, 2006; Burlingame et al., 2009). Tigoni, which is white-cream in flesh colour, had lower concentrations of carotenoids compared to yellow fleshed Dutch Robjin. As indicated by preliminary analysis, due to dehydration, carrot powder has high concentration of β-carotene

(19,217 μg/100 g) compared to 4,797 μg/100 g in fresh carrots. Addition of carrot powder onto crisps, therefore, increased the potential of crisps' contribution towards increased vitamin A intake by consumers.

Table 1. Effect of carrot powder on total carotenoid, β-carotene and reduced ascorbic acid

Cultivar	Treatment	Total carotenoids μg/100g, DWB	β-carotene μg/100g, DWB	Reduced ascorbic acid (mg/100g, DWB)
Tigoni	Raw	115.38 ± 5.11k	ND	117.28 ± 4.02ab
	Crisps0	253.84 ± 11.25h	ND	77.34 ± 1.73h
	Crisps1	472.14 ± 20.95e	89.57 ± 2.29f	92.80 ± 2.08f
	Crisps2	921.42 ± 74.24b	274.88 ± 16.36c	112.30 ± 3.83c
Cangi	Raw	132.96 ± 14.60j	0.11 ± 0.02i	128.79 ± 11.91a
	Crisps0	302.06 ± 6.43g	0.26 ± 0.07g	80.68 ± 4.44g
	Crisps1	567.87 ± 12.11d	130.38 ± 9.70c	102.86 ± 4.13de
	Crisps2	1220.49 ± 14.13a	381.89 ± 5.87a	124.04 ± 4.76a
Dutch Robjin	Raw	190.67 ± 3.17i	0.06 ± 0.03j	103.66 ± 13.07de
	Crisps0	381.33 ± 6.33f	0.21 ± 0.07h	73.34 ± 1.37i
	Crisps1	705.46 ± 11.72c	106.05 ± 7.73e	82.14 ± 1.54g
	Crisps2	1269.01 ± 78.66a	358.79 ± 17.12b	93.64 ± 1.75f

Crisps0 = without carrot powder, Crisps1 = 2.5% carrot powder, Crisps2 = 5% carrot powder, ND = not detected, DWB – dry-weight-basis.

Reduced ascorbic acid significantly ($P \leq 0.05$) varied in raw tubers and processed potato crisps depending on variety and the amount of carrot powder added. The levels in raw tubers were highest and ranged from 103.66 mg/100 g in Dutch Robjin to 128.79 mg/100 g in Cangi. These levels, however, decreased significantly ($P \leq 0.05$) in fried crisps. On addition of carrot powder, there were significant ($P \leq 0.05$) and proportional increases in reduced ascorbic acid levels in all the varieties reaching maximum of 93.6, 112.3 and 124.04 mg/100 g in Dutch Robjin, Tigoni and Cangi, respectively. Potatoes are good sources of vitamin C. The levels of reduced ascorbic acid, in this study, for the raw tubers are higher than those reported by Abong' et al. (2011a) and Hamouz et al. (2009) who reported values of 65-73 mg/100 g dry-weight and 14-24 mg/100 g fresh-weight, respectively. Reduced ascorbic acid levels are highly influenced by among other factors the variety and cultural practices (Hemavathi et al., 2010). Consumption of fortified crisps will increase reduced ascorbic acid and β-carotene (the most important source of vitamin A) intake and hence improve proportion of the daily requirement of vitamin C and vitamin A (Sanusi and Adebiyi, 2009).

Effect of carrot powder on oil content, texture and colour of potato crisps

Oil content in crisps significantly ($P \leq 0.05$) varied with potato variety ranging from 26.94% in Dutch Robjin to 32.04% in Cangi (Table 2).

Table 2. Effect of carrot powder on oil content, texture and colour parameters of potato crisps from three Kenyan varieties

				Colour measurement		
Cultivar	Treatment	Oil content (%)	Texture (N)	L*	a*	b*
Cangi	Crisps0	32.04 ± 0.74a	0.45 ± 0.07a	57.83 ± 2.77a	1.61 ± 1.36c	25.71 ± 2.22e
	Crisps1	25.89 ± 0.06e	0.43 ± 0.04a	52.83 ± 4.17bc	2.37 ± 2.29ab	30.92 ± 2.93d
	Crisps2	24.01 ± 0.03f	0.49 ± 0.05a	46.61 ± 5.77d	3.32 ± 3.08a	32.27 ± 3.67cd
Dutch Robjin	Crisps0	26.94 ± 0.17d	0.28 ± 0.03c	53.33 ± 4.34bc	-0.17 ± 0.69c	23.53 ± 1.19ef
	Crisps1	24.90 ± 0.49f	0.28 ± 0.02c	50.43 ± 2.77	2.59 ± 0.92ab	37.11 ± 2.36b
	Crisps2	23.05 ± 0.12g	0.32 ± 0.03b	47.24 ± 1.66d	3.71 ± 3.08a	39.14 ± 0.94a
Tigoni	Crisps0	30.04 ± 0.11b	0.27 ± 0.02c	52.30 ± 1.96c	-0.23 ± 0.76c	22.32 ± 1.23f
	Crisps1	27.77 ± 0.45c	0.29 ± 0.03bc	46.56 ± 2.48d	2.08 ± 0.43ab	35.48 ± 1.37bc
	Crisps2	25.86 ± 0.55e	0.31 ± 0.03b	44.25 ± 1.42de	3.37 ± 0.46a	39.15 ± 0.28a

Crisps0 = without carrot powder, Crisps1 = 2.5% carrot powder, Crisps2 = 5% carrot powder.

Table 3. Sensory properties of crisps from three Kenyan varieties flavoured with carrot powder

Cultivar	Treatment	Colour	Flavour	Texture	Oiliness	Overall
Cangi	Crisps0	5.7 ± 0.8a	5.3 ± 1.1a	5.1 ± 0.2a	5.1 ± 0.9a	5.1 ± 1.1a
	Crisps1	3.5 ± 0.6cd	4.0 ± 0.9b	4.4 ± 0.1b	4.4 ± 1.3cd	4.3 ± 0.9b
	Crisps2	3.6 ± 1.4cd	3.4 ± 1.2c	3.8 ± 0.5cd	4.0 ± 0.8d	3.6 ± 1.4c
Dutch Robjin	Crisps0	5.4 ± 1.1a	4.9 ± 1.0a	5.0 ± 0.9a	5.0 ± 0.8ab	5.0 ± 1.1a
	Crisps1	4.4 ± 0.9bc	4.1 ± 1.1b	4.8 ± 1.0ab	4.7 ± 1.0c	4.3 ± 1.1b
	Crisps2	3.4 ± 1.5cd	4.0 ± 1.4b	4.0 ± 0.8c	4.7 ± 1.1c	3.7 ± 1.2c
Tigoni	Crisps0	5.1 ± 1.0ab	4.9 ± 0.8a	5.0 ± 0.7a	4.8 ± 0.7bc	4.9 ± 0.8a
	Crisps1	4.1 ± 1.0c	4.0 ± 0.9b	4.5 ± 0.8b	4.5 ± 0.7cd	4.2 ± 0.9b
	Crisps2	3.5 ± 1.3cd	3.5 ± 0.4c	3.7 ± 0.9cd	4.0 ± 0.6d	3.7 ± 1.2c

Crisps0= without carrot powder, Crisps1=2.5% carrot powder, Crisps2=5% carrot powder. Evaluations were done on a 7-point hedonic scale; a score of 4 was the lower limit of acceptability.

Crisps oil content is negatively associated with dry matter content:- the higher the dry matter content the lower the oil content and vice versa (Abong' and Kabira, 2011). The oil content was therefore low in Dutch Robjin compared to Cangi, which had a lower dry matter content. There were significant ($P \leq 0.05$) and proportional decreases in levels of oil on addition of carrot powder reaching lowest values of 23.05%, 24.01% and 25.86% in Dutch Robjin, Cangi and Tigoni, respectively. The decrease in oil content could be accounted for by increase in total weight of crisps caused by addition of carrot powder which has trace amounts of oil.

Crisps texture (N) significantly ($P \leq 0.05$) varied with variety being high in Cangi (0.45 N) and lower in Dutch Robjin (0.28 N) and Tigoni (0.27 N). Addition of carrot powder did not have significant ($P > 0.05$) change in texture of crisps processed from Cangi. There were, however, significant ($P \leq 0.05$) increases in texture of crisps processed from varieties Tigoni and Dutch Robjin. Variety interaction with carrot powder had no significant ($P > 0.05$) influence on potato crisps texture.

The lightness colour parameter (L*) significantly ($P \leq 0$ .05) varied with variety and amount of carrot powder added on potato crisps. It was highest (57.83) in unflavoured Cangi crisps and lowest (44.25) in flavoured Tigoni crisps. The higher the levels of carrot powder, the lower the lightness value in all the varieties. There were no significant ($P > 0.05$) differences in the redness colour parameter (a*) among the varieties. There was, however, significant ($P \leq 0.05$) proportional increase in redness with addition of carrot powder reaching highest values at 5% flavouring. Yellowness parameter (b*) significantly ($P \leq 0.05$) varied with potato variety and addition of carrot powder ranging from 22.32 to 25.71 in unflavoured crisps from Tigoni and Cangi, respectively. There were significant ($P \leq 0.05$) increases in yellowness parameters on addition of carrot powder, depending on varieties. The lowest increase (32) was noted in Cangi while both Tigoni and Dutch Robjin had highest (39) increase.

Potato crisps colour and texture are important attributes that guide consumer choices and are mainly determined by frying temperature and time (Lisinska and Golubowska, 2005; Pedreschi et al., 2005) and potato variety (Abong' et al., 2011b). The effect of carrot powder on colour of crisps from all the varieties is apparent in the current study. The lightness parameter reduced on addition of carrot powder compared to significant increases in redness and yellowness parameters due to the fact that carrots have yellow and red pigments that were introduced in crisps on flavouring.

Sensory properties of potato crisps flavoured with carrot powder

All the sensory properties of crisps flavoured with carrot powder with the exception of oiliness did not significantly ($P > 0.05$) differ with potato variety (Table 3). Variety and flavour interaction with the sensory attributes with the exception of colour was equally insignificant ($P > 0.05$).

Colour scores significantly ($P \leq 0.05$) decreased with addition of carrot powder with Cangi being the most affected, all the flavoured crisps were unacceptable (3.5 and 3.6). Crisps from Tigoni (4.1) and Dutch Robjin (4.4) with 2.5% carrot powder were, however, acceptable. Flavouring using carrot powder significantly ($P \leq 0.05$) reduced sensory scores on flavour, colour, oiliness and overall acceptability. With the exception of Cangi and Tigoni crisps that were flavoured at 5%, all the other crisps were acceptable in terms of flavour and texture. Variety Dutch Robjin was least affected in flavour and texture, being acceptable at all levels of flavouring. Oiliness scores were all acceptable at any rate of flavouring as indicated by values of ≥4. Generally, all the unflavoured crisps were more acceptable to the panelists. Crisps flavoured at the rate of 2.5% were acceptable irrespective of the variety. On the other hand, all crisps flavoured at 5% were unacceptable to panelists. It therefore implies that proper awareness creation will be necessary for increased uptake of flavoured crisps to be realized.

There were no significant ($P > 0.05$) correlations between sensory texture scores and objective textural measurements. There were, however, significant ($P \leq 0.05$) correlations between sensory colour scores and the lightness (L*), redness (a*) and yellowness (b*) parameters (Table 4). Lightness parameter positively correlated to sensory colour score while redness and yellowness parameters were negatively correlated. This indicates that consumers are highly keen on colour perception and any visual change in normal colour of crisps as a result of flavouring can easily be detected leading to acceptance or rejection of the product.

Table 4. Pearson correlation coefficient (r) between sensory texture, colour scores and objective parameters of crisps made from three potato varieties

	Objective parameters			
Sensory attributes	Texture	L*	a*	b*
Texture	-0.25	N/A	N/A	N/A
Colour	N/A	0.8[a]	-0.82[b]	-0.79[a]

[a]Significant correlation coefficient (P=0.01). (N=9)
[b]Significant correlation coefficient (P=0.007). (N=9)
N/A=not applicable

CONCLUSION

As a flavouring agent, carrot powder enhanced the levels of β-carotene and ascorbic acid in potato crisps processed from three Kenyan varieties. However, scores on sensory attributes significantly decreased on application of carrot powder. Up to 2.5% carrot powder may be used as crisps flavour without adverse effect on acceptability and therefore consumption. This study did not establish the shelf-life of flavoured crisps and it would be important to know how long they can store at various environmental conditions.

ACKNOWLEDGEMENTS

Authors acknowledge the Kenya Agricultural Productivity and Agribusiness Project (KAPAP) and the National Potato Research Centre (KARI) for financial support. The role played by potato breeders, Mr. S. Nderitu and Mr. J. Onditi (NPRC), who ensured that the materials were available is acknowledged. We appreciate the Food Science Department team at KARI-Tigoni led by Mrs. N. Ngone and Mr. K. Bethuel who offered their service during sample evaluation. Laboratory analysis was accomplished with assistance from Mr. J. M'Thika and Ms. R. Kamau, Laboratory

Technicians at the University of Nairobi. The authors also recognize the support from Mr. Bonfus Nyaga of Bright Bonika Foods Ltd.

REFERENCES

Abong'GO, Okoth MW, Imungi, JK, Kabira JN (2010a). Consumption patterns, diversity and characteristics of potato crisps in Nairobi, Kenya. Journal of Applied Biosciences, 32:1942–1955.

Abong'GO, Okoth MW, Imungi, JK, Kabira JN (2010b). Characteristics of the potato crisps processing industry in Kenya. Journal of Animal and Plant Sciences, 8(1): 936- 943.

Abong'GO, Okoth MW, Imungi, JK, Kabira JN. (2011a). Losses of ascorbic acid during storage of fresh tubers, frying, packaging and storage of potato crisps from four Kenyan potato cultivars. American Journal of Food Technology, 6(9): 772-780.

Abong'GO, Okoth MW, Imungi, JK, Kabira JN (2011b). Effect of slice thickness and frying temperature on color, texture and sensory properties of crisps made from four Kenyan potato cultivars. American Journal of Food Technology, 6(9): 753-762.

Abong' GO, Kabira JN (2011). Suitability of three newly released Kenyan potato varieties for processing into crisps and French fries. African Journal of Food, Agriculture, Nutrition and Development, 11(6): 5266-5281.

Andreu-Sevilla AJ, Hartmann A, Burlo F, Poquet N, Carbonell-Barrachina A (2009). Health benefits of using red palm oil in deep-frying potatoes: Low acrolein emissions and high intake of carotenoids. Food Science and Technology International, 15: 15-22.

AOAC (1980). Official Methods of Analysis, 13th ed. Association of Official Analytical Chemists. Washington DC, USA.

Berger M, Kuchler T, Maaben A, Bush-StockfishM, Steinhart H, (2008). Correlations of carotene with sensory attributes in carrots under different storage conditions. Food Chemistry, 106: 235-240.

Bonierbale M, Gruneberg W, Amoros W, Burgos G, Salas E, Porras E, Felde TM, (2009). Total and individual carotenoid profiles in *Solanum phureja* cultivated potatoes: II. Development and application of near-infrared reflectance spectroscopy (NIRS) calibrations for germplasm characterization. Journal of Food Composition and Analysis, 22: 509–516.

Brown CR (2008). Breeding for phytonutrient enhancement of potato. American Journal of Potato Research, 85: 298–307.

Burgos G, Salas E, Amoros W, Auqui M, MunoaL, Kimura M, Bonierbale M, (2009). Total and individual carotenoids profiles in *Solanum phureja* of cultivated potatoes: I. Concentrations and relationships as determined by spectrophotometry and HPLC. Journal of Food Composition and Analysis, 22: 503–508.

Burgos G, Amoros W, Salas E, Muñoa L, Sosa P, Díaz C, Bonierbale M, (2011). Carotenoid concentrations of native Andean potatoes as affected by cooking. Food Chemistry, Doi:10.1016/j.foodchem.2011.09.002.

Burlingame B, Mouille B, Charrondiere R (2009). Nutrients, bioactive non-nutrients and anti-nutrients in potatoes. Journal of Food Composition and Analysis, 22: 494–502.

Dellapenna D, Pogson BJ (2006). Vitamin synthesis in plants: Tocopherols and carotenoids. Ann. Review of Plant Biology, 57: 711–738.

Ducreux LJM, Morris WL, Hedley T, Shepherd HV, Davis S, Millam, Taylor MA (2005). Metabolic engineering of high carotenoid potato tubers containing significant levels of B-carotene and lutein. *Journal of Experimental Biology,* 56: 81-89.

Edem, OD, (2011). Vitamin A: A review. In Asian Journal of Clinical Nutrition, 1(1): 65-82.

Ezekiel RN, Singh S, Sharma, Kaur A (2011). Beneficial phytochemicals in potato — a review. Food Research International, doi:10.1016/j.foodres.2011.04.025.

Hamouz, K, Lachman, L, Dvorak, P, Orsak, M, Hejtmankova, K, Cizek M (2009). Effect of selected factors on the content of ascorbic acid in potatoes with different tuber flesh color. Plant Soil Environment, 55(7): 281–287.

Hemavathi C, Upadhyaya P, Akula N, Young KE, Chun SC, Kim DH, Park SW (2010). Enhanced ascorbic acid accumulation in transgenic potato confers tolerance to various abiotic stresses. Biotechnology Letters, 32(2): 321-330. International Federation of Fruit Juice Products (IFFJP). (1972). Determination of total carotenoid and β-carotene. 44: 1-4.

Jansky SH (2010). Potato flavor. American Journal of Potato Research, 87(2): 209-217.

Johnston, C (2009). Functional foods as modifiers of cardiovascular disease. American Journal of Lifestyle Medicine, 3: 39S-43S.

Lisinska G, Golubowska G (2005). Structural changes of potato tissue during French fries production. Food Chemistry, 93: 681-687.

Lulai E, Orr PH (1979). The influence of specific gravity on oil content of chips. American Potato Journal, 56: 379-390.

Lung'aho C, Kabira JN (1999). A guide to growing table potatoes. Kenya Agricultural Research Institute, Nairobi, Kenya.

Hayashi M, Naknukool S, Hayakawa S, Ogawa M, Ni'matulah AA (2011). Enhancement of antimicrobial activity of a lactoperoxidase system by carrot extract and β-carotene. Food Chemistry, 130: 541–546.

Martin FL, Ames JM (2001. Comparison of flavor compounds of potato chips fried in palmolein and silicone fluid. Journal of American Oil Chemical Society, 78 (8): 863-866.
Morris WL, Ducreux L, Griffiths DW, Stewart D, Davies HV, Taylor MA (2004). Carotenogenesis during tuber development and storage in potato. Journal of Experimental Botany, 55: 975–982.
Pedreschi F, Moyano P, Granby K (2005a). Color changes and acrylamide formation in fried potato slices. Food Research International, 38: 1–9.
Pedreschi R, Betalleluz-Pallardel I, Chirinos R, Curotto C, Campos D (2011). Impact of cooking and drying on the phenolic, carotenoid contents and in vitro antioxidant capacity of Andean Arracacha (*Arracacia xanthorrhiza* Bancr.) root. Food Science and Technology International, 17(4): 319–330.
Sanusi RA, Adebiyi AE (2009). Beta-carotene content of commonly consumed foods and soups in Nigeria. Pakistan Journal of Nutrition, 8 (9): 1512-1516.
Taylor, MA (2005). Metabolic engineering of high carotenoid potato tubers containing significant levels of β-carotene and lutein. Journal of Experimental Botany, 56: 81–89.
Toledo A, Burlingame B (2006). Biodiversity and nutrition: A common path toward global food security and sustainable development. Journal of Food Composition Analysis, 19: 477–483.
Zude M, Birlouez-Aragon I, Paschold P, Douglas N, Rutledge DN (2007). Non-invasive spectrophotometric sensing of carrot quality from harvest to consumption. Post-harvest Biology and Technology. 45, 2007, p. 30–37.

8. Effects of lactic acid fermentation on the sensory profile of orange fleshed sweet potato

Benard O. Oloo[1], Anakalo Shitandi[2*], Simeon Mahungu[1], Joyce B. Malinga[3]
[1]Department of Food Science and technology, Egerton University, Egerton, Kenya.
[2]Division of Research, Kisii University, Kisii, Kenya.
[3]Kenya Agricultural Research Institute, Njoro, Kenya.
[*]*Corresponding author Email: ashitandi@kisiiuniversity.ac.ke or ashitandi@gmail.com*

SUMMARY

Lacto-products processed by lactic acid fermentation are known to have nutraceutical attributes and create variety in the beverage composition. Due to their high nutritive value, they are beneficial to human health when consumed regularly. In this study orange fleshed potato varieties (Zapallo, Nyathiodiewo and SPK004/06) were fermented with *Lactobacillus plantarum* MTCC 1407 at 25±2°C for 48 hours and kept for 28 days to make lacto-pickles. An optimization of process conditions was done by varying brine levels with fermentation time. The fermented products were subjected to panelist evaluation for flavour profiling. The product sensory scores were (1.5-2.5) on a 5-point hedonic scale ranging from dislike slightly to like much. The product with brine levels at 4 and 6% were found to be most preferred. The findings from this study indicate the acceptability of the developed lacto-pickles and suggest that the development of new products from sweet potato with functional attributes could improve its consumption among families while enhancing the shelf-life of the product. The developed lacto-pickles could also find wider acceptance with good prospects for commercialization in small-scale industries. It was concluded that the β-carotene rich, sweet potato lacto-pickle is a novel product and its regular consumption could help in combating Vitamin A deficiency related conditions such as night blindness, liver injury, aging and related ailments because of its anti-oxidant properties

INTRODUCTION

Sweet potato is an important part of the staple diet of populations in tropical regions where it is grown up to an elevation of 2,000 m above sea level (O' Sullivan et al., 1997; Owori et al., 2007). It is grown in abundance around major lakes in the East African Rift Valley of Uganda, Rwanda, Burundi, Tanzania and Kenya (Kapinga et al., 2005; Ndolo et al., 2007). Sweet potato roots are rich in starch, sugar, vitamin C, β-carotene, iron and minerals (Ray and Ward, 2006). Their leaves are high in essential amino acids: lysine and tryptophan. These two amino acids are always limiting in cereals thus, the sweet potato leaves can easily complement cereal based diets in the region (Kapinga et al., 2005).

In Africa, sweet potatoes are usually consumed without special processing. The fresh tubers are either, boiled, baked, roasted or fried. Sweet potato flour and starch may also be processed (Owori et al., 2007). Utilizing new cultivars of orange-fleshed sweet potatoes with nutritious benefits is one way to expand the market opportunities for the sweet potato industry (Suda et al., 2003). The development of new products from sweet potato with functional attributes could further improve its consumption among families while improving the shelf life of the product (Ndolo et al., 2007; Kapinga and Agili, 2007; Mwanri et al., 2011). Some varieties of sweet potatoes contain coloured pigments such as β-carotene, anthocyanin and phenolic compounds. These pigments form the basis of classifying the foods as nutraceuticals (Ray and Ward, 2006). They are considered anti-oxidants having physiological attributes such as anti-cancer, protection against night blindness, ageing and liver injury (Ray and Ward, 2006). Nutritionally, sweet potatoes usually have higher protein content than other tubers such as cassava and yams.

In this study orange-fleshed sweet potato varieties (Zapallo, Nyathiodiewo and SPK004/06) were fermented with *Lactobacillus plantarum* MTCC 1407 at 25±2°C for 48 hours and kept for 28 days to make lacto-pickles. An optimization of process conditions was done by varying brine levels with fermentation time. The new lacto-products from sweet potato had enhanced nutraceutical attributes and sensory (texture, taste, flavour and after taste) attributes of sweet potato lacto-juice were evaluated.

MATERIALS AND METHODS

The project was carried out at Egerton University, Kenya Agricultural Research Institute (KARI)-Njoro and Kenya Industrial Research Development Institute (KIRDI) laboratories. Sweet potatoes of the varieties: SPK/004/6, Zapallo and Nyathiodiewo were used for the experiment. The orange-fleshed sweet potato samples were collected randomly from KARI, Njoro, demonstration plots during peak harvesting season. KARI-Njoro is at an altitude of 2,185 m above-sea-level and lies between 0° 20° South and 35° 56° East of Kenya. The area receives an average rainfall of 939 mm per annum, with mean temperature of 14.9°C. The site is classified as lower highland 2-3 agro-ecological zone and has a sub-humid modified tropical climate. The soil type is predominantly mollic andosols (Jaetzold and Schmidt, 2007). Samples were packed in 10 kg colour-coded mesh bags and transported to the laboratory where they kept at room temperature (25 ± 2°C) in a lockable compartment prior to use within 24 hours. The experiment was laid out in a completely randomized block design. A ten day period was allowed to elapse before the second collection of replicate samples.

Preparation of *Lactobacillus plantarum* (MTCC 1407) culture

The *Lactobacillus plantarum* (MTCC 1407) culture was procured from the Institute of Microbial Technology, India. The bacterial culture was grown on MRS Agar medium to test potency and then counted to determine the viable cells. The culture was kept at 30°C for 24 hours and then extended to 36 hours.

Pickling of sweet potato

The roots were prepared by hand washing, peeling and then cutting into small cubes (approximately 1 x 1 x 1 cm). The cubes were blanched at 70°C for 10 minutes. Blanched sweet potato cubes (140 g) were dispensed in 500 ml plastic jars. Brine solution at five different concentrations (2, 4, 6, 8 and 10%) were prepared by dissolving 10, 20, 30 40 or 50 g salt in 500 ml distilled water. 300 ml of the prepared brine solution was added to each bottle. The starter culture (7 ml, 1 x10^7 cfu/ml) was inoculated into each bottle and then each bottle was capped tightly. A negative control sample (with no salt and no *Lactobacillus plantarum)* was maintained. The entries were randomized in three replications.

The sweet potato-brine mixture was fermented on the laboratory bench at room temperature. Fermentation was allowed to continue for 28 days for the pickle to season before potassium metabisulphite (100 mg/g)) was added to prevent the growth of undesirable microorganisms such as salt-tolerant fungi, *Aspergillus* species. The flow diagram for this process is shown in Figure 1.

Sensory evaluation and flavour profiling

A semi-trained panel was involved in flavour profiling of the basic attributes of the product. The product was subjected to affective sensory test according to the procedure below. Sensory attributes (taste, aroma, flavour, colour/appearance and after-taste) were evaluated using a 5-point hedonic scale (where 1 = dislike extremely and 5 = like extremely) by an 18 member panel (gender: 10 women, 8 men; age group: 20–40) selected from staff of Egerton University. The samples were served in polypropylene transparent cups, which had been labeled with a 3-digit random number. Questionnaires and water to rinse the mouth between each tasting sessions were provided. Panelists were advised to have at least 15 seconds break before tasting the next sample.

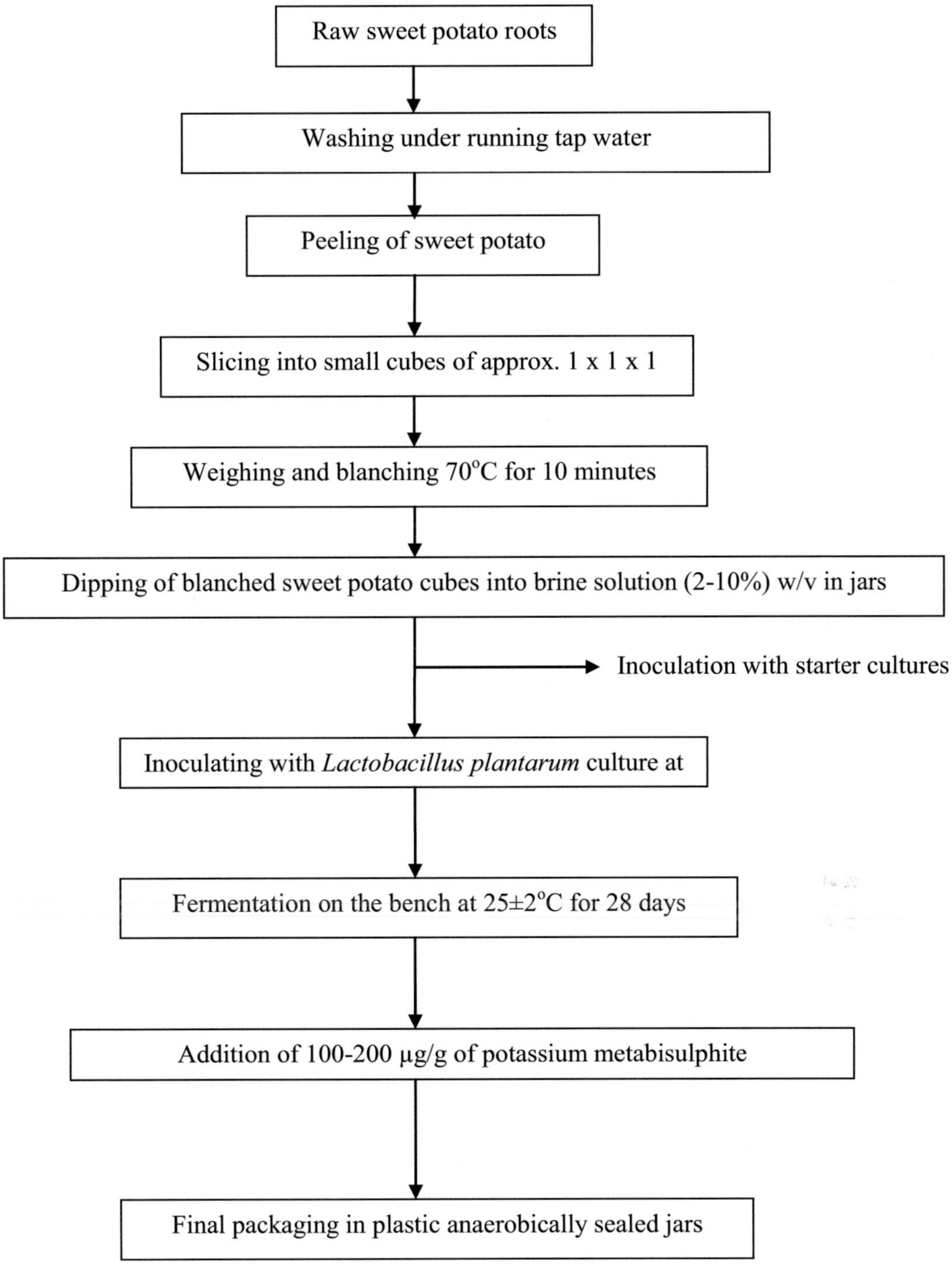

Figure 1. Flow diagram for fermentation of lacto-pickles. Adopted with modification from Panda et al., 2007).

Prior to evaluation, a session was held to familiarize the panelists with the product. The panel was then asked to read through the questionnaires and understand the meaning of each attribute (texture, taste, aroma, flavour, colour/appearance and after-taste) to avoid any misinterpretation (Kilcast and Subramanian, 2000). The tasters were not allowed to discuss their findings with one another during the evaluation session. Another set of lacto-pickles were evaluated as replication the following day. Sensory evaluation data was presented as a mean of the 18 panelists' score.

Statistical analysis

The experiment was designed in a completely randomized design and the statistical analysis was performed by using JMP Software version. One-way analysis of variance and least significant difference test were used to identify the difference between treatment means at $P \leq 0.05$.

RESULTS AND DISCUSSION

Table 1 shows an analysis of variance for the sensory evaluation of the cooked OFSP varieties. The results indicate that the Zapallo variety outperformed the others in most of the parameters under investigation and gave the highest scores on general acceptability with a mean score of 4.0. Coupled with the fact that it had the highest amount of β-carotene of 170.02 mg/kg; this variety was selected for the following fermentation process steps.

Table 1. Sensory evaluation of cooked orange fleshed sweet potato varieties

Sensory quality	Orange-fleshed sweet potato variety		
	Zapallo	Nyathiodiewo	SPK004/06
Appearance	3.9^{a}	3.6^{a}	2.5^{b}
Taste	3.9^{a}	3.5^{a}	2.7^{b}
Flavour	3.8^{a}	3.2^{b}	2.5^{c}
Starchiness	3.5^{a}	2.6^{b}	2.8^{b}
Fibrousness	3.2^{a}	3.6^{a}	3.3^{a}
General Acceptability	4.0^{a}	3.5^{b}	2.7^{c}

Values are means of panelists' scores: 1=dislike extremely; 2=like moderately; 3=like much; 4=like very much; 5=like extremely.*Means on the same row followed by the same letters are not significantly different according to least significant difference at $P > 0.05$

Table 2. The panel test results for the fermented samples

Sensory quality	Brine content					
	0%	2%	4%	6%	8%	10%
Flavour	2.14^{ab}	2.14^{ab}	2.3^{a}	2.3^{a}	1.93^{ab}	1.5^{c}
Taste	2.5^{a}	2.6^{a}	2.6^{a}	2.7^{a}	2.3^{a}	2.5^{a}
General appearance	2.5^{a}	2.5^{a}	2.2^{a}	2.14^{a}	2.0^{a}	2.3^{a}

*Means on the same row followed by the same letters are not significantly different according to least significant difference at $P > 0.05$.

Values are means of panelists' scores: 1 = dislike extremely; 2 = like moderately; 3 = like much; 4 = like very much; 5 = like extremely.

From Table 2 it was observed that for flavour, the panelists' scores ranged from 1.5-2.3 on a hedonic scale of 0 to 5. In a study conducted by Panda and Ray (2007), panelist responses with regard to flavour of lacto-pickles were reported as 3.2. This test score was however significantly different for lacto-pickles to which they added 10% sugar whose score was 4.5. The results (Table 2) suggest that the consumer responses to the lacto-pickles under this study may have been greatly enhanced by incorporating sugar. Evidently, most panelists in their comments reacted to the high salt level which may have been masked by the sugar. Furthermore, the added sugar would provide more substrate for *Lactobacillus plantarum* metabolism leading to a product of more physiological and microbial integrity.

Sensory evaluation is a unique discipline that makes use of experimental design and statistical analysis concepts to human senses with the aim of evaluating consumer products (Kuenzel et al., 2011). The main tool for sensory analysis is often descriptive based on a given scale (Algeyer and Lee, 2010). In every situation either the panelists are trained on the experiences of a particular attribute (Bayarri et al., 2011) or they may not be specifically trained. The latter is especially used where a manufacturer is only interested in whether or not their product is acceptable to their target consumer. These subjective attributes are therefore used to discriminate and distinguish products based on aroma, flavour, texture, after-taste and sound properties (Kuezel et al., 2011).

Sensory evaluation may also perform other functions such as quality control, tracking of quality changes in a product on storage and for investigation of general consumer perception of a product (Ofori et al., 2005). The resulting data helps rate products based on the chosen attributes. Correlation analyses as well as Principal Component Analysis have been used to measure linear variations between sample variables and reduce a large number of variables to two or three principal components respectively (Panda et al., 2006). The sensory analysis therefore showed the panelists score their acceptability and preference on the various selected attributes: after-taste, appearance and flavour.

From Figure 2 and 3 it was observed that panelists responded with a preference for the 4% and 6% brine on both taste and flavour. It was further evident that increasing amount of salt beyond 6% was not agreeable to most panelists (Figure 2 and 3). The 10% brine fermented samples reported the lowest scores for flavour (Figure 2). Regarding taste, the scores ranged between 2.3 and 2.7 although there were no significant differences ($P > 0.05$) among the samples (Table 2). The profile for general appearance as observed in Figure 4 showed no significant differences ($P > 0.05$) in their preference for the appearance of various sweet potato samples processed by lactic acid fermentation. The control received highest scores for appearance though, followed by 2% and the 10% brine processed lacto-pickle. This may be explained by the fact that these experienced the lowest degradation or fermentation as suggested by low level of acidity and higher accompanying pH value.

There was a general comment by consumers with regard to amount of salt. The concern for the high salt content could result to lower appreciation of the product. This however provides an avenue for improving the product by adding sugar. Addition of sucrose could influence consumer responses positively. This approach would be double pronged. First it will avail more substrate for the action of *Lactobacillus plantarum*. Second, it would lead to a quicker rise in the lactic acid content and finally, the sweetness would increase consumer acceptability. It would be of future interest to determine the exact percentage of sucrose which could be added for enhanced acceptability.

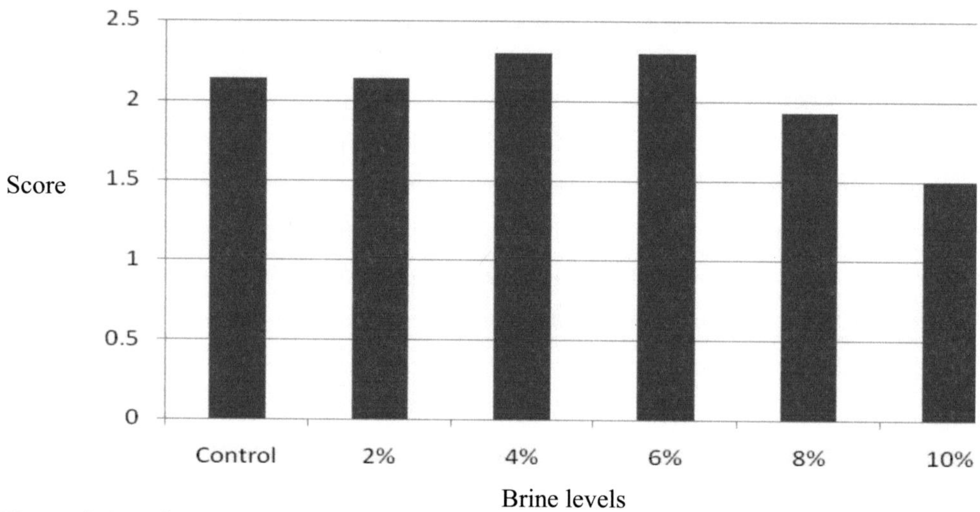

Figure 2. Panelists' response for flavour.

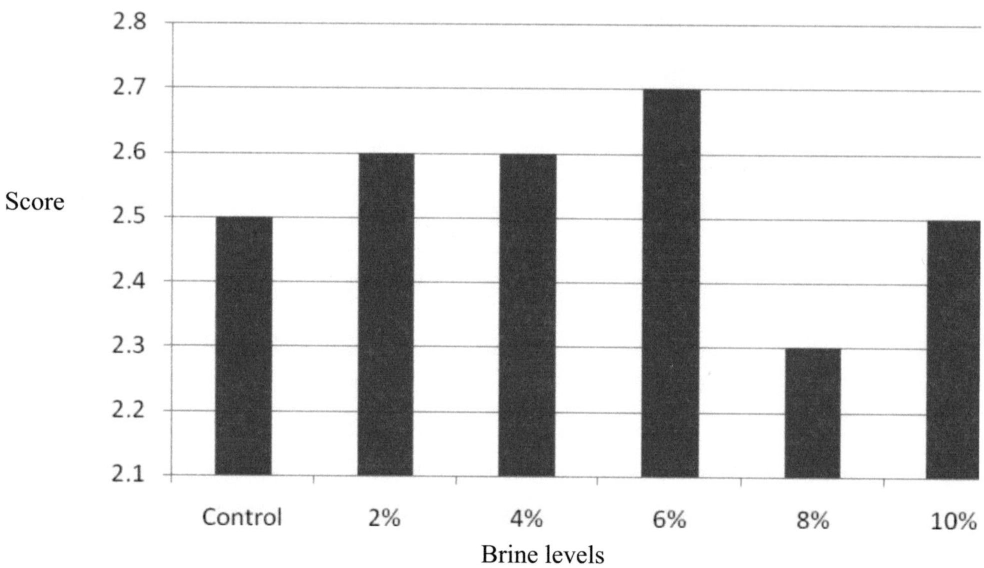

Figure 3. Panelists' response for taste.

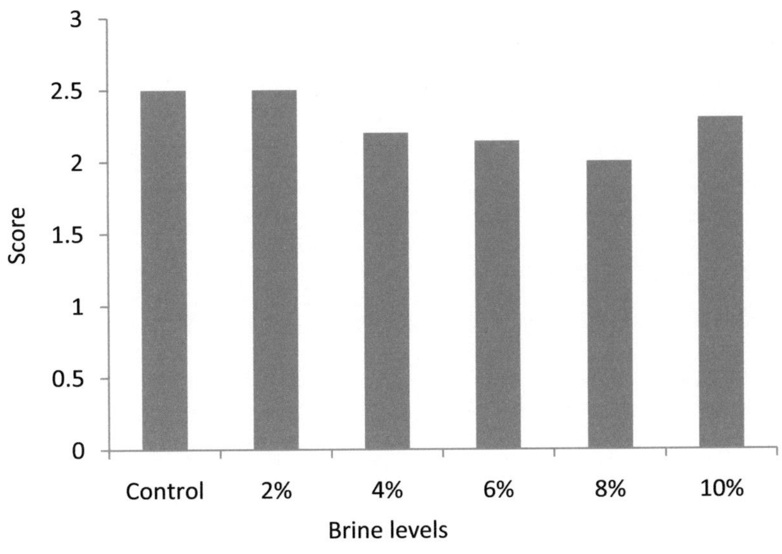

Figure 4. Panelists' response for general appearance.

CONCLUSION

The findings from this study indicate the acceptability of the developed lacto-pickles. The development of new products from sweet potato with functional attributes will improve its consumption among families while improving the shelf-life of the product. The developed lacto-pickles will also find wider acceptance with good prospects for commercialization in small-scale industries. The β-carotene rich, sweet potato lacto-pickle would be a novel product similar to lactic acid-fermented cucumber, garlic and cabbage. Its regular consumption could be helpful in combating vitamin A deficiency related conditions such as night blindness, liver injury, aging and related ailments because of its anti-oxidant properties. It would also add value to the produce, and therefore increase revenue for farmers and processors.

ACKNOWLEDGMENTS

The authors are grateful to the National Council for Science and Technology (NCST) for financial support through a research grant. We also thank Kenya Industrial Research Development Institute (KIRDI), Kenya Agricultural Research Institute (KARI) and Department of Food Science,Egerton University for kindly availing their laboratory facilities.

REFERENCES

Jaetzold R, Schmidt H. (2007).Farm Management Handbook of Kenya, Vols.II B - Central Kenya (Rift Valley and Central Provinces).Ministry of Agriculture, Livestock and Fisheries in Cooperation with GAT of GTZ, Nairobi.

Kaguongo W, Ortmann G, Wale E, Darroch M, Low J. (2012). Factors influencing adoption and intensity of adoption of orange flesh sweet potato varieties: Evidence from an extension intervention in Nyanza and Western provinces, Kenya. African Journal of Agricultural Research, 7(3):493-503.

Kapinga R, Andrade M,Lemaga B, Amanada G, CrissmanC, WwangaR. (2005). Roleof orange fleshed sweet potatoes in disaster mitigation: Experiences from East and Southern Africa. Africa Crop Science Conference Proceedings, 7: 321-1329.

Kapinga R, Tumwegamire S, Lemaga B, Andrade M, Mwanga R., Mtunda K, Ndolo P, NsumbaJ, Agili S, Serwadda B. (2005). Development of farmer based seed systems for healthy planting materials and increased sweet potato production in East and Southern Africa.

Mwanri AW, Kogi-Makau W, Laswai HS. (2011). Nutrients and antinutrients composition of raw, cooked and sun-dried sweet potato leaves. African Journal of Food Agriculture, Nutrition and Development, 11:5.

Ndolo PJ, Nungo RA, Kapinga RE, Agili S. (2007). Development and promotion of orange-fleshed sweet potato varieties in Western Kenya. In: Proceedings of the 13th ISTRC Symposium, pp. 689–695

O'Sullivan M. (2011). Analytical Methods - Proximate and Other Chemical Analyses. Encyclopedia of Dairy Sciences 2nd Edition, pp 76-82.

O'Sullivan L, Ross RP, Hill C. (2002). Review: Potential of bacteriocin producing lactic acid bacteria for improvements improvements in food safety and quality. Biochimie 84:593-604.

Owori CB, Lemaga ROM, Mwanga R, Namutebi A, Kapinga R. (2007). Sweetpotato recipe book: sweetpotato processed products from Eastern and Central Africa. African Crop Science Society, Kampala, Uganda. 93 pp.

Ray RC, Ward OP. (2006). Post-harvest microbial biotechnology of tropical root and tuber crops. In Microbial Biotechnology in Horticulture (Ray RC, Ward OP, eds.) 1:345–395, Science Publishers Inc., Enfield, NH.

Ranganna S. (2001). Proximate Analysis, Color Measurement and Sensory Evaluation. Tata McGraw Hill Co. Ltd., New Delhi, India.

Suda I, Oki T, Masuda M, Kobayashi M, Nishiba Y, Furuta S. (2003). Review: physiological functionality of purple-fleshed sweet potatoes containing anthocyanins and their utilization in foods. JARQ 37(3):167–73.

9. The micro-nutrient content in coloured varieties of sweet potatoes (*Ipomea batatas*) grown in Vihiga county, Kenya

Mildred Nawiri[1*], Kennedy Away[1] and Hudson Nyambaka[1]
[1]Chemistry Department, Kenyatta University.
**Corresponding author: mildredoyugi@yahoo.com*

SUMMARY

Nutrient deficiencies are a major global health problem affecting more than 2 billion people in low-income countries. Deficiencies are reported for both organic nutrients such as vitamin C, beta-carotene (the vitamin A precursor) and for mineral nutrients such as calcium, iron and potassium. In Kenya, potential nutrient-rich foods such as *Ipomea batatas* (sweet potato) are locally available but their nutrient content needs to be quantified. Vihiga County grows white, yellow, purple and orange coloured flesh sweet potato varieties. This study aimed at quantifying vitamin A and C in raw and boiled varieties of sweet potato from Vihiga County. In addition, raw samples of sweet potato as well as soils growing them were quantified for calcium, iron, potassium and copper. Liquid chromatography, iodometric titration and atomic absorption spectroscopy were employed. The orange fleshed varieties had significantly higher level of iron, potassium, copper and vitamins. Vitamin content reduced significantly with boiling. The level of calcium, iron, potassium and copper in soils was positively correlated ($r = 0.07$) with those found in the sweet potato. This study indicates that the orange fleshed sweet potato contain higher levels of iron, copper, potassium, and vitamin A and C making it a rich source for both organic and mineral dietary nutrient. While production of other sweet potato varieties is encouraged, the study findings promote the consumption of orange fleshed sweet potato to address nutrient deficiencies.

INTRODUCTION

Deficiencies of nutrients are still a major global health problem with more than 2 billion people mostly in low income countries estimated to be deficient (World Health Report, 2004). Deficiencies are reported for organic nutrients such as vitamin C and beta-carotene, which is a vitamin A precursor. Also reported are deficiencies in mineral nutrients including calcium, iron and potassium. These deficiencies increase the general risk of both infectious and non-communicable illnesses and death from diarrhoea, measles, malaria and pneumonia (World Health Report, 2004). Pregnant women, lactating women and young children, are most vulnerable mainly because they have a relatively greater need for vitamins and minerals (Nabakwe and Ngare, 2004; Ministry of Health, 2010). Children in particular have an increased risk of vitamin A deficiency which is globally a leading cause of disease burden and mortality. Deficiencies have been pegged to lack of access to nutrient-rich foods such as fruits, vegetables, animal products and fortified foods partially because they are expensive or locally unavailable. The importance of food-based approaches for the prevention and control of nutrient deficiencies as well as for the improvement of nutrition in general is fully recognized worldwide (FAO, 2005; Musinguzi et al., 2006; Nawiri et al., 2012).

Sweet potato (*Ipomoea batatas* (L) Lam) belongs to the botanical family Convolvulaceae. It is amongst the world's most important, versatile and under-exploited food crops with more than 133 million tons in annual production. It is the fifth most important food crop on fresh-weight basis in developing countries after rice, wheat, maize and cassava and is cultivated in over 100 developing countries (CIP, 1999). While the tuber is consumed as a staple food, other roles are reported and as well other parts of the crop have been found to be important. Other than being combined with lime juice to make dye for clothes, sweet potatoes have also been used as fodder for animals and as green vegetables (Woolfe, 1992; Abidin, 2004). The skin colours of sweet potatoes range from white, cream, yellow, orange, pink, red or purple. Flesh colours may be white or various shade of cream, yellow, orange or even purple (Purseglove, 1991; Woolfe, 1992; Abidin, 2004; Abubakar et al., 2010).

The desirable nutritional value of *Ipomoea batatas* is gaining recognition, as the understanding between diet and health increases. Various parts of the crop have been reported to contain both organic and mineral nutrients including vitamins A and C, zinc, potassium, sodium, manganese, calcium, magnesium and iron (Woolfe, 1992; Leighton, 2007; Antia et al., 2006; Oduro et al., 2008; Ofori et al., 2009; Abubakar et al., 2010; Meludu, 2010; Ingabire and Vasanthakaalam, 2011; Ukom et al., 2011; Hue et al., 2012). According to Woolfe (1992), potassium is the major mineral in sweet potatoes followed by phosphorous, calcium, magnesium, iron and copper.

The nutrient content in plants vary depending on differences in soil mineral content, interaction between minerals, varieties and processing (Landon, 1991; Woolfe, 1992; Ravindran et al., 1995; Kruger et al., 1998; Chattopadhyay et al., 2006; Leighton, 2007; Abubakar et al., 2010; Sunette, 2010; Ingabire and Vasanthakaalam, 2011; Ukom et al., 2011). Okum and co-workers (2009) found a significant increase in the mineral contents of most sweet potato varieties with application of nitrogen fertilizer. Toxicity of copper in plants is rare but soil concentrations above 100 mg/kg are regarded as high while a range of 0.19-0.50 mg/100 g in sweet potatoes has been reported (Woolfe, 1992; Kruger et al., 1998). Potassium though is readily available in the soil solution for uptake by plant roots. Calcium is very important in plant growth and is not considered toxic to plants but high levels of iron and especially potassium may reduce its uptake in crops. Studies report calcium in sweet potatoes to vary in the range of 7-85 mg/100 g while level of iron varies between 0.16-2.11 mg/100 g (Woolfe, 1992; Abubakar et al., 2010; Sunette, 2010). The white and cream fleshed sweet potatoes have been found to virtually contain no beta-carotene while the range in coloured fleshed varieties is less than 100 μg to more than 26,600 μg/100 g (Laurie, 2001; Van Jaarsveld et al., 2005; Van Jaarsveld et al., 2006). However, the retention of beta-carotene and vitamin C decreases due to either boiling, sun drying or storage (Van Jaarsveld et al., 2006; Bengtssonet al., 2008; Bechoff et al., 2009; Nawiri et al., 2012).

Vihiga County, one of the most densely populated areas in western Kenya, lies between longitudes 34° 30′ and 35° 0′ East and between latitudes 0° and 0° 15′ North, with four districts Emuhaya, Vihiga, Hamisi and Sabatia. The region receives an annual rainfall ranging from 1,740-1,940 mm with average temperature ranging from 14-34°C in January and 14-30°C in July (District Strategic Plan, 2005). Orange, purple, white and yellow fleshed varieties of sweet potatoes are grown throughout the year in this county but high prevalence of nutrient-related diseases have also been reported (Akelola et al., 2007). While the desirable nutritional qualities place sweet potato in a unique position in the quest to eliminate nutrition deficiencies in communities of Vihiga County, literature on the nutrient content of the sweet potatoes is scarce.

MATERIALS AND METHODS

Sampling and sample pretreatment

Three farmers were identified in each of the four districts of Vihiga county: Emuhaya, Vihiga, Hamisi and Sabatia. Under regular practice of planting sweet potatoes (i.e. without use of fertilizer) each farmer planted the white, yellow, orange and purple varieties of sweet potatoes in their farms (experimental plots of 3 m x 3 m). About 200 g of each variety was randomly sampled thrice after 90 days of growth at intervals of 10 days up to 120 days, as per the farmer's practice of harvest (Woolfe, 1992). The different varieties of sweet potatoes were packed in separate paper bags and transported to the laboratory where they were washed to remove soil debris. The outer cover was peeled off before being cut into small pieces and samples of the same coloured variety obtained during the three sampling sessions homogenized before analysis. Samples to be analyzed in their raw form for mineral nutrients were dried at 70°C for 24 hours to moisture content of less than 10% before digestion. Otherwise they were boiled at 30°C for 15 minutes before storage. Samples for analysis of vitamin A and C in the raw form were stored fresh at -4°C. Concurrently during sweet potatoes sampling, 500 g of soil was randomly sampled from each plot and packed in separate paper bags for transportation to the laboratory where they were dried at 70°C for 24 hours to a moisture content of less than 10% before digestion.

Chemicals

Potassium hydroxide, sodium sulphite, soluble starch, potassium iodate salt, potassium iodide copper nitrate, iron (II) nitrate, butylated hydroxytoluene (BHT), potassium chloride, all-trans-BC (Type IV), vitamin C solid, calcium carbonate, lanthanum chloride were sourced from Sigma-Aldrich through the local distributor (Kobian Chemicals Ltd). All other chemicals used were of analytical grade.

Sample digestion and measurement of mineral nutrients

Standard procedures were followed according to Ukom *et al* (2011). Briefly, about 10 g (in triplicates) of dried sweet potato was ground in a mortar and sieved through a 0.2 mm sieve. The powder (1 g) was wet digested with ultra pure nitric acid and digested for 10 minutes at 120^0C while adding few drops of hydrogen peroxide. The solution was filtered and diluted with deionised water before aspiration into the spectrophotometer. Triplicate soil samples were also subjected to similar treatment. Atomic absorption spectrophotometer (model 210 VGP, Buck Scientific Inc, Connecticut, USA) was used for the measurement of copper, iron and calcium while a Sherwood 410 flame photometry (Sherwood Scientific Ltd, Cambridge, UK) was used for analysis of potassium. The operating parameters were set according to the specification given by the manufacturer including lamp current of 1.2 amperes and fuel system of air/acetylene. Instrumental calibration was done with blanks with the following ranges of working standards: 1.0-5.0 ppm calcium, 0.1-0.5 ppm copper, 0.5-2.5 ppm iron and 2-20 ppm potassium. The equation of each generated calibration curve was used in calculating analyte concentrations in the soils and sweet potato samples.

Extraction and measurement of vitamins

Beta-carotene extraction procedure was modified according to Nawiri et al. (2012). Briefly, sweet potato samples previously stored at -4°C were thawed then 3 g were mashed in a mortar. To this was added 50 ml 0.5 M ethanolic potassium hydroxide, 5.0 ml 10% sodium sulphite solution (w/v) and 0.3 g ascorbic acid. n-Hexane was added to the mixture, shaken thoroughly and allowed to settle before decanting the hexane layer into a separating funnel. The residue was re-extracted four times and the combined extracts were washed with 1 M potassium hydroxide, followed by equal portions of distilled water until decolourized. Hexane was dried by filtering over anhydrous sodium sulphate and evaporated at 40°C in a rotatory evaporator to almost dryness while breaking the vacuum with nitrogen gas. The residue was then re-dissolved in 5 ml ethanol, filtered through a 0.45 µm millipore and stored at -4°C before injection into the high performance liquid chromatography (HPLC) column.

A Shimadzu SCL-10AVP HPLC system (Shimadzu Corporation, Tokyo, Japan) was calibrated using freshly prepared beta-carotene standards (calibration range 2-10 ppm) with the absorbance set at 450 nm. Sample extracts were checked for the UV-Vis maximum absorbance before injection into the HPLC column to avoid column overloading. Sample extracts (20 µl) were injected into the column through 20 µl injection loops using a micro-syringe (1750 Gastight model, Hamilton Company, USA). Separation was performed isocratically with a degassed mobile phase consisting of methanol/DCM/water in the ratio 80:18:2 (v:v:v) containing 0.1% butylated hydroxytoluene on a C_{18} column (250 X 4..6 mm id, 5 µm particle size, Vydac TP-201) with absorbance units set between 0.4-1.2 AU. The elution took 8 minutes at a flow rate of 1.0 ml/min. Beta-carotene was monitored using a UV-Vis detector (tungsten lamp) at 450 nm at sensitivity of 0.0500 absorbance unit's full scale. All trans-BC was identified by comparing the retention times with that of standard solutions while peak areas were used for quantification.

For vitamin C analysis, 10 ml distilled water was added to 10 g sweet potatoes sample and mashed in a mortar. The extract was decanted and the process repeated twice. The sample extract was strained through a cheese cloth rinsing the pulp with distilled water into a 100 ml volumetric flask and topping up to the mark using distilled water. The extract was titrated with standardized iodine solution.

Data analysis

Data were analyzed with SPSS 17.0 for windows. The mean and standard deviation of means were calculated. The data were analyzed by one-way analysis of variance and Duncan's Multiple Range Test was used to separate means ($P \leq 0.05$).

RESULTS AND DISCUSSION

Mineral nutrients in soils

The content of copper, iron, potassium and calcium in soils from four districts in Vihiga county are presented in Table 1.

Table 1. Mineral content (mean ± sd) of soils growing sweet potatoes in districts of Vihiga County.

Element	Site	Mineral content (mg/100g)	P-value
Copper	Sabatia	1.525 ± 0.01^{d} *(1.5-1.53)*	0.000
	Hamisi	1.25 ± 0.01^{b} *(1.25-1.28)*	
	Emuhaya	1.375 ± 0.01^{c} *(1.35-1.40)*	
	Vihiga	0.85 ± 0.01^{a} *(0.83-0.88)*	
Iron	Sabatia	217.85 ± 2.60^{b} *(211.88 219.69)*	0.000
	Hamisi	296.53 ± 3.23^{d} *(292.87-298.98)*	
	Emuhaya	285.29 ± 1.53^{c} *(283.98-286.98)*	
	Vihiga	225.78 ± 2.26^{a} *(223.46-227.98)*	
Potassium	Sabatia	73.9 ± 1.00^{a} *(71.25-76.25)*	0.000
	Hamisi	100.5 ± 0.13^{d} *(99.68-100.25)*	
	Emuhaya	91.05 ± 2.60^{b} *(83.98-96.75)*	
	Vihiga	93.68 ± 1.75^{c} *(90.00-98.50)*	
Calcium	Sabatia	34.70 ± 0.20^{b} *(34.13-35.00)*	0.000
	Hamisi	35.43 ± 0.34^{c} *(34.45-35.98)*	
	Emuhaya	30.68 ± 0.32^{a} *(30.00-31.58)*	
	Vihiga	39.58 ± 0.18^{d} *(39.25-40.00)*	

Range in parenthesis.
Same small letters for a specific element in the column are not significantly different at $P > 0.05$.

The levels of mineral nutrients in the soils from all districts decreased in the order iron > potassium > calcium > copper. Copper levels ranged between 1.25-1.53 mg/100 g in the county with the soils in Sabatia having significantly higher level than the other three districts (P = 0.000). The findings however, indicate safe levels of copper in the soils of Vihiga county (Landon, 1991). Iron levels in soil were found to be high with a range between 564.45 mg/100 g and 741.33 mg/100 g with Hamisi district recording significantly higher levels than the other districts (P = 0.000). Potassium levels in soil ranged between 73.90-100.50 mg/100 g being significantly high in Hamisi district (P = 0.000). Levels of potassium are attributed to the readily available potassium found in soil solution. For calcium, significantly higher levels were recorded in Vihiga district as compared to the other districts (P = 0.000). The calcium levels in the county are however possibly reduced by the high levels of other cations such as iron and potassium which are also known to reduce its uptake in crops.

Mineral content in sweet potatoes

The content of copper, iron, potassium and calcium in the white, purple, orange and yellow varieties of sweet potatoes from the four districts of Vihiga county are presented in Table 2. Generally the levels of mineral nutrients in sweet potatoes from the districts decreased in the order calcium > potassium > iron > copper indicating a linear positive correlation (r = 0.700) between the mineral nutrients in sweet potato and in the soils. These findings are in line with the work of previous authors (Woolfe, 1992; Antia et al., 2006; Ukom et al., 2011).

The content of copper in varieties of sweet potatoes ranged between 0.25-0.65 mg/100 g within the county. Toxicity of copper rarely occur naturally in soils and this range of copper in Vihiga County was described as safe (Landon, 1991). In nutrition however, copper works with iron in the formation of red blood cells as well as keeps the blood vessels, nerves, immune system, and bones healthy. Although Antia and co-workers (2006) report that copper was completely absent in sweet potato leaves, our findings are within ranges of other studies (Woolfe, 1992; Kruger et al., 1998). The orange fleshed sweet potatoes had significantly higher copper content irrespective of the district they were grown ($P \leq 0.05$). In comparison to other coloured varieties of sweet potatoes, this variety has generally higher levels of nutrients (Laurie, 2001; Van Jaarsveld et al., 2006).

The iron content in sweet potatoes ranged between 1.03-1.40 mg/100 g in the county this being attributed to the high levels of iron in the soil from Vihiga County (Table 1). While the levels found in this study are in agreement with earlier findings, higher levels of iron in sweet potatoes have also been attributed to differences in soil mineral content and interaction between minerals and varieties (Woolfe, 1992; Kruger et al., 1998; Abubakar et al., 2010; Meludu, 2010; Sunette, 2010; Ingabire and Vasanthakaalam 2011; Ukom et al., 2011). The potential benefits of the orange fleshed variety are also observed here as the findings show that its iron content was higher in comparison to the other varieties.

Table 2. Mean of mineral nutrients (mg/100g) in varieties of sweet potatoes grown in districts of Vihiga county.

		Districts in Vihiga County							
		Sabatia		Hamisi		Emuhaya		Vihiga	
Mineral	Sweet potato variety	Mean ± sd[1]	P-value	Mean ± sd[1]	P- value	Mean ± sd[1]	P-value	Mean ± sd[1]	P-value
Copper	White	0.45±0.01[b]	0.003	0.35±0.02[b]	<0.001	0.33±0.02[a]	0.000	0.25±0.01[a]	0.000
	Purple	0.35±0.0[a]		0.53±0.02[c]		0.45±0.01[b]		0.45±0.01[b]	
	Yellow	0.45±0.01[b]		0.28±0.01[a]		0.38±0.01[a]		0.48±0.01[b]	
	Orange	0.50±0.02[b]		0.65±0.01[d]		0.60±0.01[c]		0.55±0.01[c]	
Iron	White	1.30±0.01[b]	0.001	1.30±0.01[a]	0.004	1.28±0.02[a]	0.870	1.10±0.02[a]	<0.001
	Purple	1.28±0.02[b]		1.28±0.01[a]		1.30±0.01[a]		1.30±0.02[b]	
	Yellow	1.12±0.02[a]		1.28±0.02[a]		1.28±0.02[a]		1.03±0.02[a]	
	Orange	1.38±0.02[b]		1.40±0.02[b]		1.28±0.02[a]		1.33±0.02[b]	
Potassium	White	10.40±0.20[a]	0.000	14.66±1.04[a]	<0.001	16.50±0.38[a]	0.000	8.38±0.13[a]	<0.001
	Purple	16.43±0.32[c]		18.25±0.54[b]		13.25±0.55[a]		13.70±0.34[b]	
	Yellow	13.43±0.27[b]		14.25±0.15[a]		16.38±0.48[a]		14.60±0.23[b]	
	Orange	131.78±1.09[d]		136.98±1.22[d]		168.38±2.10[b]		149.85±1.78[c]	
Calcium	White	25.88±0.37[b]	0.001	25.30±0.23[c]	<0.001	26.30±0.11[c]	0.000	25.93±0.52[b]	<0.001
	Purple	24.43±0.03[a]		18.50±0.22[a]		23.70±0.43[b]		24.13±0.14[b]	
	Yellow	27.35±0.02[c]		26.30±0.01[d]		24.75±0.10[b]		25.78±0.57[b]	
	Orange	24.31±0.31[a]		22.00±0.25[b]		22.02±0.23[a]		21.28±0.07[a]	

[1]Same small letters for a specific element in the column are not significantly different at $P > 0.05$.

The potassium content in varieties of sweet potatoes in the county ranged between 10.4-168.38 mg/100 g. Studies have reported levels of potassium that are above 129 mg/100 g in sweet potatoes attributing that to application of fertilizers (Woolfe, 1992; Ravindran et al., 1995; Ukom et al., 2011). However, no fertilizer was applied to plots growing the sweet potatoes in this study. It is worth noting that for potassium, the orange fleshed variety grown in all districts also had significantly higher content ($P \leq 0.05$) in comparison to other varieties.

Calcium content in varieties of sweet potatoes in Vihiga county ranged between 18.50-27.35 mg/100g probably being lower due to the high levels of iron and potassium which reduce calcium uptake in crops. Although the levels found in this study are either higher or lower than previous studies, calcium is not known to be toxic to plants (Woolfe, 1992; Ravindran et al., 1995; Sunette, 2010). Contrary to copper, iron and potassium being highest in the orange variety, different varieties were found to contain higher calcium in Vihiga county. The yellow fleshed variety grown in Sabatia and Hamisi and the white variety grown in Emuhaya had significantly higher calcium content ($P \leq 0.000$). While the reason may not be explained within the scope of this study, this could be attributed to factors mentioned that cause differences in nutrient content of plants (Woolfe, 1992; Ravindran et al., 1995; Kruger et al., 1998; Leighton, 2007; Abubakar et al., 2010; Meludu, 2010; Sunette, 2010; Ingabire and Vasanthakaalam 2011; Ukom et al., 2011).

Beta-carotene and vitamin C content in sweet potatoes

The content of beta-carotene and vitamin C in the four varieties of sweet potatoes grown in Vihiga county are presented in Table 3. In raw sweet potatoes, beta-carotene was detected in the yellow and orange fleshed varieties only, therefore, supporting studies that found virtually no beta-carotene in white fleshed varieties (Van Jaarsveld et al., 2005; Ingabire and Vasanthakaalam, 2011). In the county, the range of beta-carotene was 4,619- 4,889 μg/100 g in the orange variety being significantly higher than 2,017-2,628 μg/100 g of the yellow fleshed variety ($P \leq 0.05$). The findings support previous studies on the significant role of orange fleshed sweet potatoes to be utilized as a viable food-based strategy for controlling vitamin A deficiency (Laurie, 2001; Van Jaarsveld et al., 2006). Boiling, however, resulted in losses of beta-carotene (Table 3). Beta-carotene is known to be susceptible to degradation upon exposure to heat, light and oxygen (Van Jaarsveld et al., 2006; Bengtsson et al., 2008; Nawiri et al., 2012).

The sweet potato varieties in Vihiga county contained vitamin C in the range 4.85-5.73 mg/100 g but boiling significantly reduced this to about 0.88 mg/100 g. Various processing procedures including baking, frying and boiling are attributed to degradation of this vitamin (Lee and Kader, 2000; K'Babalola et al., 2010). The raw orange fleshed variety had significantly higher ($P \leq 0.05$) content of vitamin C than the other varieties indicating its potential to provide for this vitamin.

Table 3. Mean content of beta-carotene (µg/100g) and vitamin C (mg/100g) in raw and boiled varieties of sweet potatoes grown in districts of Vihiga County.

Districts	Swet potato variety	Beta-carotene (raw)		Beta-carotene (boiled)	Vitamin C (raw)		Vitamin C (boiled)	
		Mean±sd	P-value	Mean±sd	Mean±sd	P-value	Mean±sd	P-value
Sabatia	White	ND		ND	4.85 ± 0.00^{a}		0.81±0.13	
	Purple	ND		ND	5.51 ± 0.00^{b}		0.88±0.01	
	Yellow	2628 ± 2.50^{a}	0.003	ND	5.51 ± 0.00^{b}		0.81±0.13	
	Orange	4853 ± 2.63^{b}		ND	5.51 ± 0.38^{b}	0.0049	0.88±0.01	0.596
Hamisi	White	ND		ND	4.85 ± 0.00^{a}		0.88±0.01	
	Purple	ND		ND	5.30 ± 0.13^{b}		0.81±0.13	
	Yellow	2452 ± 2.54^{a}	0.004	ND	5.58 ± 0.13^{b}		0.88±0.01	
	Orange	4778 ± 1.78^{b}		ND	5.58 ± 0.13^{b}	0.000	0.88±0.00	0.441
Emuhaya	White	ND		ND	4.85 ± 0.01^{a}		0.81±0.13	
	Purple	ND		ND	5.36 ± 0.13^{a}		0.88±0.01	
	Yellow	2267 ± 1.46^{a}	0.002	ND	5.58 ± 0.13^{b}		0.81±0.13	
	Orange	4889 ± 2.56^{b}		ND	5.73 ± 0.01^{a}	0.000	0.73±0.13	0.487
Vihiga	White	ND		ND	4.85 ± 0.01^{a}		0.88±0.01	
	Purple	ND		ND	5.51 ± 0.01^{b}		0.88±0.01	
	Yellow	2071 ± 2.47^{a}	0.004	ND	5.51±0.01		0.88±0.01	
	Orange	4619 ± 2.74^{b}		ND	5.66±0.13	0.000	0.88±0.01	-

[1]Same small letters for a specific element in the column are not significantly different at $P \leq 0.05$.

ND; no data.

CONCLUSION

This study indicates that the orange fleshed sweet potatoes contain high levels of iron, copper, potassium and vitamin A and vitamin C making it a very rich source for both organic and mineral dietary nutrients. However, the other varieties can contribute to nutrient requirements of children and are strongly recommended.

ACKNOWLEDGEMENT

The authors are grateful to Jomo Kenyatta University of Agriculture and Technology, Department of Foods and Nutrition for the availing their HPLC and to Department of Geology and Mines, for the mineral analysis.

REFERENCES

Abidin PE. (2004). Sweet potato breeding for North Eastern Uganda: Farmer Varieties, farmer participatory, selection and stability of performance. Wageningen University, Netherlands, PhD thesis.

Abubakar HN, Olayiwola IO, Sanni SA, Idowa MA. (2010). Chemical composition of sweet potatoes (Ipomoea batatas L am) dishes as consumed in Kware state, Nigeria. International Food Research Journal, 17: 411- 416.

Akelola R, Karuri NM, Mwaniki D. (2007). The effect of orange fleshed sweet potato on vitamin A nutritional status of pre-school children in Nambale Division, Busia District. Proceeding of the 13th ISTRC Symposium 13:721-727.

Antia BS, Akpan EJ, Okon PA, Umoren IU. (2006). Nutritive and antinutritive evaluation of sweet potato leaves. Journal of Nutrition, 5(2):166-168.

Bengtsson A, Namutebi A, Larsson AM, Svanberg U. (2008). Effects of various traditional processing methods on all the trans-β-carotene content of orange fleshed sweet potatoes. Journal of Food Composition and Analysis, 21:134-143.

Chattopadhyay A, Chakraborty I, Mukhopadhyua SK, Kumar PR, Sen H. (2006). Compositional changes of sweet potato as influenced by cultivation, harvest date and cooking. Journal of Horticulture, 705:211-218.

CIP (International Potato Centre). (1999). CIP sweet potato: Production, Utilization, Consumption, Feed use. Downloaded from http://www.cipotato.org.

District Strategic Plan. 2005. Implementation of the national population policy for sustainable development. National coordinating Agency for population and development, Nairobi, Kenya.

FAO. 2005. Food and agricultural organization of the United Nations: Major food and agricultural commodities and procedures. Downloaded from http://www.fao.org/es/ess/top/commodity.

Hue SM, Boyce AN, Chandran S. (2012). Antioxidant activity, phenolic and flavonoid contents in different varieties of sweet potato (ipomea batatas). Australian Journal of Crop Science, 6(3):375-380.

Ingabire MR, Vasanthakaalam H. (2011). Comparison of the Nutrient Composition of four sweet potatoes varieties cultivated in Rwanda. American Journal of Food and Nutrition doi:10.5251

K' Babalola OO, Adubiaro HO, Ikusika O. (2010). The effect of some processing methods on vitamin C content of sweet potatoes and Irish potatoes. Electronic Journal of Environmental, Agricultural and Food Chemistry, 9(4):679-681.

Kruger M, Sayed S, Langenhoven M, Holing F. (1998). Composition of South African Foods, Vegetables and Fruits. Supplement to the MRC Food composition Tables 1991.Tygerberg: Medical research council, Pretoria, South Africa, 245 – 278.

Landon JR. (ed) (1991). Booker Tropical Soil Manual; John Wiley and Sons Inc.; New York.

Laurie SM. (2001). Overview of breeding and evaluation of orange fleshed sweet potato in South Africa. Proc. of VITAA project regional workshop, Nairobi, Kenya.

Lee SK, Kader AA. (2000). Preharvest and postharvest factors influencing vitamin C content of horticultural crops. Postharvest Biology and Technology, 20:207-220.

Leighton CS. (2007). Nutrient and sensory quality of Orange fleshed Sweet potatoes. Msc. Thesis, University of Pretoria, South Africa.

Meludu NT. (2010). Proximate analysis of Sweet potato toasted granules. African Journal of Biomedicine 13:89-91.

MOH. (Ministry of Health). (2010). Outpatient cover (MOH 204B)-June-September records, Register 5, Republic of Kenya, Mbale District Hospital, Kenya.

Musinguzi E, Kikafunda JK, Kiremire BT. (2006). Utilization of indigenous food plants in Uganda: A case study of south –western Uganda. African Journal of Food and Agriculture Nutrition and Development, 6(2):723-724.

Nabakwe EC, Ngare KD. (2004). Health and nutritional status of children in Western Kenya in relation to Vitamin A deficiency. East African Journal of Public Health, 1:1-8.

Nawiri MP, Nyambaka HN, Murungi JI. (2012). Sundried cowpeas and amaranthus leaves recipe improves β-carotene and retinol in serum and hemoglobin concentration among preschool children. European Journal of Nutrition doi 10.1007/s00394-012-036-2.

Oduro I, Ellis WO, Owusu. (2008). Nutritional potential of two leafy vegetables: *Moringa oleifera* and *Ipomea batatas* leaves. Scientific Research and Essay, 3:57-60.

Ofori G, Oduro I, Ellis WO, Dapaah H. (2009). Assessment of Vitamin A content and sensory attributes of new sweet potato (ipomea batas) genotypes in Ghana. African Journal of Food Science 3(7):184-192.

Purseglove JW. (1991). Tropical crops dicotyledons. Longman Scientific and Technical. John Wiley and sons, 1st ed., New York, USA.

Ravindran V, Ravindran G, Sivakenesan R, Rajaguru SB. (1995). Biochemical and Nutritional assessment of tubers from 16 cultivars of sweet potato. Journal of Agricultural Food Chemistry, 43:2646-2651.

Sunette ML. (2010). Agronomic Performance, Consumer acceptability and nutrient content of new sweet potato varieties in South Africa. PhD thesis, University of Free state, South Africa, 13-56.

Ukom AN, Ojimelukwe PC, Alamu EO. (2011). All trans-cis β-carotene content of selected sweet potato (Ipomea batatas (L) Lam) varieties as influenced by different levels of nitrogen fertilizer application. African Journal of Food Science, 5(3):131-137.

Van Jaarsveld PJ, Faber M, Tanumihardio SA, Nestel P, Lombard CJ, Benadé AJS. (2005). Beta-carotene-rich orange-fleshed sweet potato improves the vitamin A status of primary school children assessed with the modified-relative-dose-response test. American Journal of Clinical Nutrition, 81(5):1080-7.

Van Jaarsveld PJ, Marais D, Harmse E, Nestel P, Rodrigues-Amaya DB. (2006). Retention of β-carotene in boiled, mashed orange-fleshed sweet potato. Journal of Food Composition and Analysis, 19(4): 321-329.

Woolfe JA. (1992). Sweet potato: An untapped food resource, Cambridge University Press and International Potato Centre (CIP), Cambridge, UK 634-643.

World Health Report. (2004). Joint statement by the World Health Organization, the World Food Programme and the United Nations Children's Fund. Preventing and controlling micronutrient deficiencies in populations affected by an emergency. Multiple vitamin and mineral supplements for pregnant and lactating women and for children aged 6 to 59 months.

10. Consumer acceptability of *gari* in coastal Kenya

Alice M. Nderitu [1], Abdul K. Faraj [2*], Prisca J. Tuitoek [1] and Rahab W. Muinga [3]
[1]Department of Human Nutrition and Preclinical Studies, Egerton University, Njoro, Kenya.
[2]Department of Dairy and Food Science and Technology, Egerton University, Njoro, Kenya.
[3]Kenya Agricultural Research Institute, Mtwapa, Kenya.
[*]*Corresponding author: ab_faraj@hotmail.com*

SUMMARY

Gari is a granular meal made by fermenting grated cassava, then toasting in a shallow pan until it becomes crisp. The fermentation step in the preparation of *gari* imparts a characteristic aroma and sour flavour which may in turn affect acceptability of the product. Hence, the consumer acceptability of *gari* fermented for various durations (2, 4 and 6 days) was studied. Clone 1838 (renamed Karembo variety) grown at KARI-Mtwapa, which is low in cyanide (< 40 ppm) was used. For all the attributes of the *gari* that were evaluated (colour, texture, aroma, flavour and general acceptability), *gari* fermented for 2 days was more acceptable than that fermented for a longer time (4 and 6 days). However, the unfermented sample was liked most in all the attributes. Flavour was found to be the single most important factor in determining the general acceptability of *gari* ($r = 0.75$). During the sensory evaluation, most of the panelists (> 90%) said that all the *gari* samples were tasteless and not appealing. Some suggested that addition of sugar and food colour could possibly make the product better. The use of other additives could thus be tried in Kenya in efforts to increase the acceptability of *gari*.

INTRODUCTION

The coastal region of Kenya has chronic food insecurity and high poverty levels necessitating recurring food relief (KDDP, 1997-2001). Cassava which is an important food crop in the region has the potential to improve food security in this area. The region produces 30% of the country's total cassava production. Researchers at Kenya Agricultural Research Institute (KARI), Mtwapa, have been involved in developing improved cassava varieties that are high yielding and tolerant to cassava pests and diseases. In efforts to increase cassava production, six new varieties were released by the Ministry of Agriculture, Livestock and Fisheries in 2008 (Muli et al., 2008). However, fresh cassava roots cannot be stored for long because they rot within 3-4 days after harvest. They are also bulky with about 70% moisture content. Hence, transportation of the roots from rural to urban markets is very difficult and expensive (Nweke, 1996). According to KARI Annual Report (KARI, 2003), a baseline survey on cassava production, usage and marketing conducted in coastal Kenya showed that there was little processing, except for cassava chips and crisps. Processors identified lack of know-how on diversification of cassava products as one of the main constraints. Thus, there is need to adopt value addition technologies that will help to make products that are less bulky and have a longer shelf-life.

Many different processing techniques are employed to convert raw cassava tubers into various products like granules, flours, pastes and chips with a range of flavours and appearances for different markets (Nweke, 1996). *Gari* is a granular meal made by fermenting grated cassava, then toasting in a shallow pan until it becomes crisp. It contributes up to 60% of the total calorie intake in West Africa (Wheatly and Chuzel, 1993). In Nigeria, trading in *gari* has led to significant improvements in the post harvest handling systems and a tremendous increase in market demand for fresh cassava roots (Wren, 1997). In *gari* making, the fermentation step is crucial to the development of the characteristic aroma and sour flavour of *gari*, imparted mainly by lactic acid bacteria, which produce lactic acid and volatiles such as aldehydes, diacetyl esters and ethanol (Wheatley and Chuzel, 1993). The duration of fermentation thus affects the consumer acceptability of *gari*. For example, commercial *gari* processors in Nigeria ferment cassava for different durations, depending on the market: up to 3 days for the Igbo people of the eastern states, who do not like a strong acid taste, but 5 days or more for the Yoruba people in the west (Nweke, 1996). Fermentation also helps to detoxify the mash, while grating allows linamarasc to comc into contact with the cyanoglucoside substrates (Wheatley and Chuzel, 1993). In Kenya, information on acceptability of *gari* is scanty. This study therefore aimed at testing the consumer acceptability of *gari* fermented at various durations (2, 4, and 6 days).

MATERIALS AND METHODS

Preparation of *gari*

Fresh cassava tubers (clone 1838, renamed Karembo variety) were obtained from KARI-Mtwapa. They were peeled, washed and grated manually to obtain cassava mash. The mash was thoroughly mixed, put in a clean sack and pressed using a manual press until no more slurry was dripping. The pressed mash was divided into four batches. Three of these were put in separate polythene bags and left to ferment naturally for 2, 4 or 6 days. The fermented mash was sifted to remove fibrous materials and "garified" (simultaneous cooking and drying) in a shallow pan while constantly stirring to prevent burning, until crispy. The fourth batch was treated like the other three except that it was not fermented. The *gari* was cooled, sieved and packed in polythene bags and sealed.

Sensory evaluation and acceptability of gari

Sensory evaluation of *gari* was done by 54 untrained panelists at KARI-Mtwapa. The panelists evaluated the *gari* for colour, aroma, flavour, texture and general acceptability on a 7-point hedonic scale which ranged from 'like very much' through 'neither like nor dislike' to 'dislike very much'. The samples were presented in identical containers, coded with 3-digit random numbers. Each sample had a different number and the sample order was randomized for each panelist. Clean water at room temperature was provided to panelists so that they could rinse their mouths before and between samples.

Chemical analysis

The most acceptable *gari* samples were analyzed for moisture, crude fat, crude protein, total ash and crude fibre using standard procedures recommended by Association of Official Analytical Chemists (AOAC, 1995). Carbohydrate content was determined by difference. The residual cyanogens in *gari* were determined using the Picrate Paper Kit method (Bradbury et al., 1999). All determinations were done in triplicates except for crude fat which was done in duplicate.

Statistical analysis

The categories on the 7-point hedonic scale were converted to numerical scores ranging from 1 to 7 (1 = dislike very much, 2 = dislike moderately, 3 = dislike slightly, 4 = neither like nor dislike, 5 = like slightly, 6 = like moderately, 7 = like very much. The numerical scores for each sample were tabulated and analyzed by analysis of variance (ANOVA) using the Statistical Analysis System (SAS, 2003). For proximate analysis data, the t-test was used to compare the differences between two means.

RESULTS AND DISCUSSION

Sensory evaluation and acceptability of gari

For all the attributes of the *gari* that were evaluated, the *gari* fermented for 2 days was more acceptable than that fermented for a longer time (4 and 6 days) as shown in Table 1. However, the unfermented sample was liked most in all the attributes. Flavour was the single most important factor in determining the general acceptability of *gari* ($r = 0.75$). The general trend showed that panelists preferred a shorter fermentation period for *gari*, or better still, no fermentation at all. In West Africa where consumption of *gari* is common, the fermentation time varies from 3 to 5 days depending on the target market (Nweke, 1996).

During the sensory evaluation, most of the panelists (> 90%) said that all the *gari* samples were tasteless and not appealing. Some suggested that addition of sugar and food colour could possibly make the product better. In Nigeria where *gari* is the dominant cassava product, it is consumed without any added ingredients or side dishes or with a variety of ingredients or side dishes such as sugar, groundnut, fish and meat. The use of other additives could thus be tried in Kenya in efforts to increase the acceptability of *gari.* The unfermented product is similar to *farinha,* a major dietary staple in Brazil (Wheatly and Chuzel, 1993). Unlike *gari,* however, typical *farinha* preparation does not include a fermentation step. The preparation proceeds directly from grating through pressing and sieving to toasting. It may be eaten dry, mixed with hot or cold water to make a paste or gruel or mixed with other foods (FAO, 1990).

Table 1. Sensory evaluation scores (mean ±sd) for the various attributes of *gari*.

Fermentation time (days)	Colour	Texture	Flavour	Aroma	General acceptability
0	$6.11^{a}\pm1.02$	5.69 ± 1.18	$5.81^{a}\pm1.45$	$5.83^{a}\pm1.16$	$5.96^{a}\pm1.20$
2	$5.68^{b}\pm0.84$	5.65 ± 1.07	$5.48^{a}\pm1.33$	$5.69^{a}\pm1.04$	$5.57^{b}\pm1.13$
4	$5.68^{b}\pm1.11$	5.57 ± 1.18	$4.94^{b}\pm1.52$	$4.98^{b}\pm1.07$	$5.17^{c}\pm1.13$
6	$5.53^{b}\pm1.04$	5.20 ± 1.38	$4.89^{b}\pm1.48$	$5.09^{b}\pm1.36$	$5.15^{c}\pm1.25$
$LSD_{0.05}$	0.34	NS	0.38	0.33	0.34

Means in a column followed by the same superscript are not significantly different at $\alpha = 0.05$.
NS - Not significant.

Proximate composition of *gari*

The unfermented product and the *gari* fermented for two days were not significantly different ($P > 0.05$) in protein, fibre, ash and carbohydrate contents but were significantly different ($P \leq 0.05$) in moisture, fat and cyanide contents (Table 2). Fermented *gari* had slightly higher protein content (1.23%) as compared to the unfermented one (1.09%). Studies have shown that the protein content of cassava increases with fermentation period (Onjoro, 1995; Tivana et al. 2007; Asegbeloyin and Onyimonyi, 2007). The study by Tivana et al. (2007) showed that the protein content of cassava flour increased from 1.3% to 1.8% (w/w) dry-weight-basis after three days fermentation while that of Asegbeloyin and Onyimonyi (2007) showed that the crude protein content of fermented *gari* was higher than that of an unfermented sample.

Table 2. Proximate composition of *gari* (% dry-weight-basis).

Component	*Gari* (unfermented)	*Gari* (fermented for 2 days)	P –Value*
Moisture	9.65 ±0.23	10.45 ±0.22	0.012
Protein	1.09 ±0.21	1.23 ±0.22	0.465
Fibre	2.16 ±0.10	2.08 ±0.03	0.258
Ash	1.33 ±0.07	1.46 ±0.05	0.053
Fat	0.75 ±0.04	0.48 ±0.04	0.020
Carbohydrate	94.56 ±0.03	94.63 ±0.08	0.384
Cyanide (ppm)	5.41 ±0.26	12.23 ±1.80	0.003

*P-value is based on the t-test.

Fermented *gari* had slightly lower fibre content (2.08%) than the unfermented sample (2.16%). This could be as a result of breakdown of complex carbohydrates into simple carbohydrates by the fermenting microorganisms. The microorganisms may also have utilized fats for their nourishment thus reducing the amount of fat present in the fermented *gari* as suggested by Onjoro (1995). Palm oil is sometimes used during toasting of *gari* to prevent the pulp from burning (Nweke, 1996). This procedure increases the lipid content of *gari* and changes the colour of the product from white to yellow. In coastal Kenya, coconut is one of the major crops grown and could be used to increase the fat content of *gari* and thus improve its caloric content and acceptability. Laswai et al. (2006) incorporated coconut in the formulation of 'kibabu' (a traditional, plain deep fried cassava product prepared from cassava paste). This not only increased the fat and protein content of 'kibabu', but also significantly improved its palatability.

Unfermented *gari* had a significantly ($P \leq 0.05$) lower cyanide content (5.41 ppm) than fermented *gari* (12.23 ppm). This could be explained by the fact that fermentation is normally accompanied by acidic conditions and reduction in pH (Wheatly and Chuzel, 1993). The cyanohydrin intermediary is quite stable in acidic conditions and linamarase activity is also reduced. Cyanohydrins may thus have accumulated in the fermenting mash and persisted even in the final product. As for the unfermented control, acidic conditions did not develop and the cyanohydrin intermediate rapidly broke down to the free cyanide.

CONCLUSION

Gari fermented for 2 days was more acceptable than that fermented for 4 or 6 days. Further, the unfermented sample was more acceptable compared to *gari* fermented for 2 days. In efforts to promote consumption of *gari* in Kenya, the use of flavour enhancers such as sugar and fat is recommended.

ACKNOWLEDGEMENTS

The authors thank Dr. J. W. Kamau (KARI-Katumani), Dr. H. M. Saha (Pwani University) and Mr. F. N. Pole (KARI-Matuga) for their valuable contribution. The financial support from the Rockefeller Foundation through grant No. FS 2006 057 for dissemination, promotion and maintenance of new cassava varieties in Kenya is acknowledged.

REFERENCES

AOAC (1995). Official Methods of Analysis.Association of Official Analytical Chemists, Washington, DC.

Asegbeloyin JN, Onyimonyi AE. (2007). The effect of different processing methods on the residual cyanide content of *gari*. Pakistan. Journal of Nutrition, 6(2): 163-166.

Bradbury MG, Egan SV, Bradbury JH. (1999). Picrate paper kits for determination of total cyanogens in cassava roots and all forms of cyanogens in cassava products. Journal of the Science of Food and Agriculture, 79:593-601.

FAO (1990). Roots, tubers, plantains and bananas in human nutrition. FAO Food and Nutrition series, No. 24. FAO, Rome.

KARI (2003). Baseline survey on cassava production, usage and marketing in coastal Kenya. In: KARI Annual Report, 2003. Pp. 42.

Kilifi District Development Plan (KDDP, 1997-2001).Office of the Vice President and Ministry of Planning and National Development.Government Printer, Nairobi, Kenya. Pp. 55.

Laswai HS, Silayo VCK, Mpagalile JJ, Balegu WR, John J. (2006). Improvement and popularization of diversified cassava products for income generation and food security: A case study of Kibabu. Journal of Food, Agriculture Nutrition and Development. Vol. 6 No. 1:1-15.

Muli BM, Saha HM, Gethi JG. (2008). New cassava varieties for farmers in coastal lowland Kenya. Farmers Pride, September Issue.

Nweke FI. (1996). Cassava processing in Sub-Saharan Africa: Implications for expanding cassava production. IITA Research, No.12: 7-14.

Onjoro PA. (1995). Effects of fermented whole cassava meal on the performance of broilers. MSc. Thesis, Department of Animal Science, Egerton University, pp. 42-45.

SAS (2003). SAS User's Guide.SAS Institute Inc., Cary NC.

Tivana LD, Bvochora J, Mutukumira AN, Owens JD, Zvauya R. (2007). Heap fermentation of cassava (Manihot esculenta Crantz) in Nampula Province, Mozambique. In: Tropical Root and Tuber Crops: Opportunities for Poverty Alleviation and Sustainable Livelihoods in Developing Countries. Kapinga R, Kingamkono R, Msabaha M, Ndunguru J, Lemaga B, Tusiime G. (eds.) Proceedings of the 13th ISTRC Symposium held at AICC Arusha, Tanzania, Pp. 445-450.

Wheatley CC, Chuzel G. (1993). Cassava. In: Encyclopaedia of Food Science, Food Technology and Nutrition. Macrae R, Robinson RK, Sadler MJ. (eds.). Academic Press Limited, London. Pp. 734-743.

Wren C. (1997). Cashing in on cassava. Post harvest systems. The Newsletter for Post Harvest Systems Development in Africa. No. 1:11-12.

11. Development of solar-biomass hybrid tunnel dryer for drying grains

Jackis Aukah[1*], Daniel Kerich[1], Julius Ogola[1]
[1]KIRDI Western Region Centre (KWRC), Kenya Industrial Research and Development Institute (KIRDI), Kisumu.
[*]*Corresponding author: aukajakiz@gmail.com*

INTRODUCTION

The high moisture content at which grains are harvested necessitates the use of dryers to remove excess moisture in order to minimize losses and quality deterioration. Commercial dryers have extensively been used to dry grain. In most current commercial crop dryers, heat for the drying process is generated by combustion of oil, gas, wood or by electrical means. Such drying methods are unaffordable to small-scale farmers in rural areas as the fuels and dryers are costly and hence not readily available to them. Furthermore the exhaust gas emitted from combustion of most fuels is often dirty and may be toxic. Solar grain dryers have been used with success for drying grain with considerable reduction in grain losses as opposed to floor or field sun drying. Solar dryers depend on the intensity of solar radiations to heat up the air and effect drying. They offer numerous advantages such as protection of drying material from contamination, quick drying and superior milling and germination qualities. However, the solar drying process is interrupted during cloudy or rainy periods and at night. Thus a combination of mixed mode dryer using solar as main input and biomass heat exchanger system as an auxiliary source of thermal energy can compensate for the absence of solar radiation.

This innovation is aimed at reducing grain wastage and improving their storage conditions with the specific objectives of designing and developing a dryer using residual heat from biomass to pre-heat air and a turbo-ventilator to create necessary draught and maintain good airflow through the dryer tunnel. The report describes a detailed design procedure and outlines a systematic combination of the application of basic design concepts and rules of thumb resulting from numerous experimental studies. The design of the dryer was based on drying 87 kg of grains by mass from initial moisture content of 35% wet-basis to a moisture content of 14% wet-basis, from which 21.5 kg of water is required to be removed by solar radiation or biomass heat exchanger or both. The drying time was estimated to take 3 days when operating on solar mode and 12 hours when operating on hybrid mode to dry the grains to desired moisture content. A mathematical model was developed to calculate the geometries of the biomass heat exchanger and tunnel dryer. The model is based on the energy balance, convection heat transfer correlations, air temperature, relative humidity of air and suitable initial assumptions. The material selection criterion is also presented since it is crucial for the success of the design. The paper also highlights a step-by-step production process of the dryer components. The fabricated hybrid tunnel dryer consists of a flat plate air heating collector and a drying tunnel fabricated as one unit. The top surfaces of the collector and drying chamber were designed in triangular shape with an inclination of 24^0 on both sides in order to increase the area of irradiance. Both the collector and the drying units were covered with UV-stabilized plastic sheet. A biomass stove-heat exchanger system was placed at the inlet of the tunnel dryer to improve the reliability of the dryer and the quality of the dried product. The heat exchanger was of cross-flow type with flue gas from the stove passing through the tubes and ambient air flowing over the tubes before entering the dryer. A turbo-ventilator was used to force ambient air into the dryer at the rate of 0.65 m^3/s. The heat exchanger was fitted with biomass stove and chimney at the opposite sides through rectangular ducts.

MATERIALS AND METHODS

Material selection of the solar–biomass hybrid tunnel dryer

Materials selection was done by considering compatibility between construction materials and working fluids, in particular with regard to corrosion and/or operation at elevated temperatures. Physical and chemical properties of the materials were also taken into account. The selected materials for the solar-biomass hybrid tunnel dryer were:

1) Hardwood for the base of the dryer because it is a poor conductor of heat and hence heat loss from the cabinet will be greatly minimized.
2) The inside of the cabinet was lined with aluminium sheet in order to reflect heat back to the cabinet and prevent decaying of the wood due to humid air.
3) Galvanized sheet metal was chosen because of its toughness and ability to conduct and radiate heat. It was used to fabricate the heat exchanger.
4) Mild steel square tube and angle line has great strength and can be easily welded. It was used to construct the main frame of the dryer.
5) Black pipe was installed in the roof of the tunnel dryer because of its strength and weldability.
6) Wire mesh was used to construct the trays because it resists corrosion and allows air to pass through it.
7) Aluminium angle line was used to make the tray supports because of its high resistance to corrosion.
8) Rock wool was used to insulate the dryer and the heat exchanger.
9) Castable refractory cement was used to make the refractory lining of the biomass stove and heat exchanger because of its ability to resist high temperatures and also poor conductor of heat.
10) Fire bricks were was chosen because they are poor conductors of heat and resistant to high temperatures. They were used to build the refractory lining of the biomass heat exchanger system.
11) UV-treated sheet were used as glazing material because they are cheap and resist photo degradation.
12) Aluminium pipe were chosen for heat transfer surface (heat exchanger tubes) because of low cost, light weight, high conductivity and good joining characteristics.

Design calculations

Solar dryer sizing

The main requirement of the system was to dry/dehydrate grains. The initial and final properties of the product vary from type to type, but for the system sizing, the following design parameters were selected:

Capacity: 87 kg of fresh product.

Initial moisture content: 35% on wet-basis.

Final moisture content: 14% on wet-basis.

Drying time: 3 days.

Design constraints

1) Ambient air temperature in the environment where the dryer was to be placed is assumed to be 26^0C.
2) Simulation studies showed that as the ratio of drying chamber area to the collector surface area was increased, the overall drying efficiency increases. However, it was observed that for the ratio in the range 1-2 there was no observable difference in the performance of the dryer. A value of 1 is therefore recommended as the ratio of the drying floor area to the collector surface area.

Based on the above data, the following parameters could be estimated:

Area of solar collector

It was assumed that the grain will be dried within 3 days (i.e. 8 active solar hours per day). The area of the solar collector was designed by considering the following parameters:

1) Mass of water to be removed from the product, M_W
2) Specific latent heat of vaporization of water from crops ($L = 2335\, kJ/kg$)
3) The solar insolation per unit horizontal area per day (I) in Kenya ranges from 4-6 kWh/m^2. Taking conservative average as 5 kWh/m^2 (18 MJ/m^2) per day to be used for design
4) The efficiency of the collector was , $\eta_c = 25\%$

The collector area was computed using the following formula;

$$A_c = \frac{M_w L}{I \eta_c N_d}$$

Where: N_d = number of days required for drying

A_c = collector aperture area

Mass of water to be removed from the product is calculated by;

$$M_w = M\left(\frac{m_i - m_f}{1 - m_f}\right)$$

Where: M = dryer capacity per batch (87 kg per batch)

m_i = initial moisture content of product to be dried (35% wet basis)

m_f = maximum desired final moisture content (14%)

Therefore:

$$M_w = 87\left(\frac{0.35-0.14}{1-0.14}\right)$$

$$M_w = 87\left(\frac{0.21}{0.86}\right)$$

$$\Rightarrow M_w = 21.2\ kg$$

Substituting the values in equation (1) above we have

$$A_c = \frac{21.2\times 2335}{0.25\times 18{,}000\times 3}$$

$$\Rightarrow A_c = 3.67\ m^2$$

Let L_c = length of the of the collector

and b_d = Breadth of tunnel dryer (taken to be equal to standard width of mild steel plate i.e.1.22 m used as collector for ease of fabrication).

Therefore, the length of the collector was given by:

$$L_c = \frac{A_c}{b_d} = \frac{3.67}{1.22}$$

$$\Rightarrow L_c = 3\ m$$

Area of drying bed

Let A_{db} = area of drying bed

Following design constraint (2) which recommends that the ratio of drying floor area to collector surface area is taken as 1, we have:

$$\frac{A_{db}}{A_c} = 1$$

$$A_{db} = A_c \times 3.67m^2$$

$$\Rightarrow A_{db} = 3.67\ m^2$$

Thus the length of the drying bed was 3 m.

Quantity of air required to effect drying in kg (Q_a)

Consider energy balance for drying which states that energy available from the quantity and temperature of air going through the dryer should be equal to the energy needed to evaporate the amount of water to be removed from the crop. Mathematically it is expressed as:

$$M_a C_p (T_d - T_i) = M_w L$$

Where: M_a = mass of dry air

C_p = specific heat capacity of air = 1.02 kJ/kgK

T_i = ambient temperature = 26°C

T_d = temperature of the dryer = 70°C

L = latent heat of vaporization of water from grain = 2335 kJ/kg

M_w = mass of water to be removed from the product = 21.2 kg

Therefore the mass of dry air was calculated as:

$$M_a = \frac{M_w L}{C_p(T_d - T_i)} = \frac{21.2 \times 2335}{1.02(70-26)}$$
$$\Rightarrow M_a = 1103\,kg$$

Volume of air to effect drying in m³ (Va)

Assumption: consider air as an ideal gas.

Air is usually quantified as volume at atmospheric pressure and temperature. Consider the equation of state for an ideal gas as shown below:

$$PV_a = M_a R T_i$$

Where: P = atmospheric pressure (101.3 kPa)

V_a = volume of dry air, m^3

T_i = atmospheric (ambient temperature)

M_a = mass of dry air (calculated as 1103 kg)

R = constant factor, it equals to about 0.291 kPa m^3/kg K

Rearranging we have:

$$V_a = \frac{M_a R T_i}{P} = \frac{1103 \times 0.291 \times 299}{101.3}$$
$$\Rightarrow V_a = 948\,m^3$$

Heat exchanger design

The design of the biomass system was based on the following considerations:

1) The heating will be indirect, i.e. flue gas from the biomass stove and the drying air will not be mixed. This will protect the product being dried from contamination by smoke, soot and ash of the flue gas.
2) The temperature of air heated by the heat exchanger, entering the dryer, would be in the range of 65-70°C. This is based on the allowable maximum drying temperature for most staple food crops.
3) Temperature control of the drying air would be possible, by controlling combustion in the stove through opening or closing the primary air supply gate in the stove.
4) Biomass operation can be carried out for extended periods of time, unattended. The stove was designed to operate continuously for about two hours for a single fuel loading, with briquetted rice husk as fuel.

Tube layout

Staggered triangular tube arrangement was chosen because it has better heat transfer and surface area per unit length (greatest tube density) and also provides a more compact arrangement, usually resulting in smaller shell and the strongest header sheet for a specified shell-side flow area.

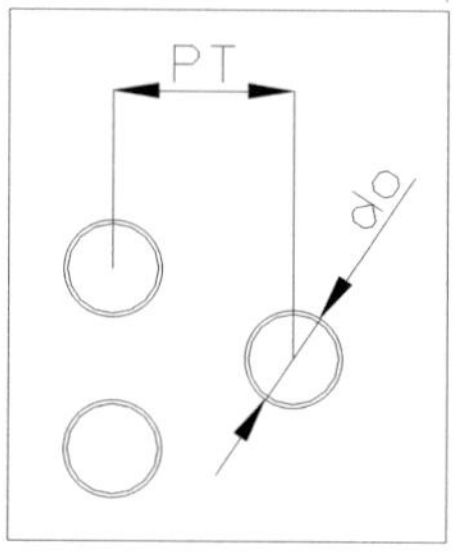

Figure 1. Staggered triangular tube arrangement.

Pitch of the tube

Tube pitch P_T is chosen so that the pitch ratio is $1.25 \le \frac{P_T}{d_o} \le 1.5$. When the tubes are too close to each other ($\frac{P_T}{d_o}$ less than 1.25), the header plate (tube sheet) becomes too weak for proper rolling of the tubes and cause leaky joints. Taking maximum value for calculation then we have:

$$\frac{P_T}{d_o} = 1.5$$
$$P_T = 1.5 d_o = 1.5 \times 50$$
$$\Rightarrow P_T = 75\ mm$$

The following assumptions were made:

The heat exchanger is at steady state.

The overall heat transfer coefficient U is constant.

The heat exchanger is perfectly insulated (i.e. heat loss to the surroundings is negligible).

There is no heat conduction in the direction of the metal separating the fluids nor in the fluids themselves.

There is a uniform temperature in each section of the heat exchanger and complete mixing takes place just before the fluids exit each passage.

The specific heat capacities are constant through the heat exchanger.

Kinetics and potential energy changes are negligible.

Exit temperature of the flue gas

Basic input parameters were:

1) Ambient air temperature, $T_{a1} = 26^0 C$

2) Maximum desired temperature of air in the dryer , $T_{a2} = 70^0 C$

3) Specific heat capacity of air , $C_{pa} = 1.005\ kJ/kgK$

4) Density of air $\rho_a = 1.285\ kg/m^3$

5) Mass flow rate of air, $\dot{m}_a = 0.27\ kg/s$

6) Flue gas inlet temperature, $T_{f1} = 600^0 C$

7) Specific heat capacity of flue gas, $C_{pf} = 1.122\ kJ/kgK$

8) Mass flow rate of flue gas is taken as, $\dot{m}_f = 0.04\ kg/s$

Expressing heat load in terms of energy balance (energy taken up by air equals to energy given up by the flue gas).

Mathematically this can be expressed as:

$$Q_L = \dot{m}_a C_{pa}(T_{a2} - T_{a1}) = \dot{m}_f C_{pf}(T_{f1} - T_{f2})$$
$$Q_L = 0.27 \times 1.005 \times (70 - 26) = 0.0355 \times 1.122 \times (600 - T_{f2})$$
$$\Rightarrow T_{f2} = 300^0 C$$

Where: T_{f2} = exit temperature of flue gas

Q_L = Heat load in kW

Heat transfer surface area

Known parameters were:

Diameter of tube, $d_o = 50\ mm$

Thickness of the tube, $b = 2\ mm$

Thermal conductivity of alluminium, $k_{al} = 204\ W/mK$

Thermal conductivity of flue gas, $k_f = 0.03524\ W/mK$

Dynamic viscosity of flue gas, $\mu_f = 2.876 \times 10^{-6}\ kg/sm$

Prandtl number, $P_r = 0.65$

Note that the properties of flue gas values were taken at flue gas exit temperature

Consider general equation for heat transfer in heat exchangers given below

$$Q_L = UA_s \Delta T_m$$
$$A_s = \frac{Q_L}{U \Delta T_m}$$

Where: U = overall heat transfer coefficient, W/m^2K

A_s = heat transfer surface area, m^2

ΔT_m = log mean temperature difference, K

Overall heat transfer coefficient can be evaluated using the following relation;

$$\frac{1}{U}=\frac{1}{h_a}+\frac{b}{k_t}+\frac{1}{h_f}$$

Where: h_a = convection heat transfer coefficient for air, W/m^2K

h_f = convection heat transfer coefficient for flue gas, W/m^2K

b = thickness of the heat exchanger tube, m

k_{al} = thermal conductivity of heat exchanger tube for aluminium, W/mK

Convection heat transfer coefficient of flue gas for turbulent flow inside the tube was determined from Dittus and Boelter equation given by:

$$N_u = 0.023\, R_e^{0.8} P_r^{0.4}$$

Note that N_u, R_e and P_r are dimensionless numbers expressed mathematically as:

Nusselt number: $N_u = \dfrac{h_f d_i}{k_f}$

Reynolds number: $R_e = \dfrac{G_f d_i}{\mu_f}$

Prandtl number: $P_r = \dfrac{C_{pf}\mu_f}{k_f}$ but this can be determined from properties tables

Where: h_f = convection heat transfer coefficient of flue gas

d_i = internal diameter of tube,

k_f = thermal conductivity of flue gas

μ_f = dynamic viscosity

C_{pf} = specific heat capacity of flue gas

G_f = mass flux of flue gas through chimney (assuming mass flux through the tubes equals mass flux through chimney) and is given as; $G_f = \dfrac{\dot{m}_f}{A_f}$

Where A_f is taken as cross sectional area of chimney given as

$$A_f = 0.27 \times 0.18 = 0.0486\, m^2 \text{. Hence } G_f = 14.32\ kg/m^2s$$

Therefore Reynolds number was given by:

$$R_e = \frac{G_f d_i}{\mu_f} = \frac{14.32 \times 0.046}{2.876 \times 10^{-6}}$$

$$\Rightarrow R_e = 229{,}040$$

Since flue gas is in the turbulent flow regime then the Dittus and Boelter equation is applicable:

$$N_u = 0.023\, R_e^{\,0.8} P_r^{\,0.4}$$

$$N_u = 0.023(229040)^{0.8} \times 0.65^{0.4}$$

$$\Rightarrow N_u = 375.68$$

Therefore convection heat transfer coefficient was given by;

$$h_f = \frac{N_u \times k_f}{d_i} = \frac{375.68 \times 0.03524}{0.046}$$

$$\Rightarrow h_f = 287.80\ W/m^2K$$

Convection heat transfer coefficient of air across a bundle of tubes can be determined from:

$$N_u = K.R_e^{\,0.6} P_r^{\,0.33}$$

Note that N_u, R_e and P_r are dimensionless numbers expressed mathematically as:

Nusselt number; $N_u = \frac{h_a d_o}{k_a}$

Reynolds number; $R_e = \frac{G_a d_o}{\mu_a}$

Prandtl number; $P_r = \frac{C_{pa}\mu_a}{k_a}$ but this can be determined from properties tables

Where: h_a = convection heat transfer coefficient of air

d_o = External diameter of tube,

K = Arrangement of tube = 0.32 for staggered arrangement

k_a = Thermal conductivity of air

μ_a = Dynamic viscosity of air

C_{pa} = Specific heat capacity of air

G_a = mass flux of air through the dryer and is: $G_a = \frac{\dot{m}_a}{A_a}$

Where A_a is taken as cross sectional area of air inlet given as

$A_a = 1.22 \times 0.24 = 0.2928\, m^2$. Hence $G_a = 0.922\ kg/m^2 s$

Therefore Reynolds number was given by:

$$R_e = \frac{G_a d_o}{\mu_a} = \frac{0.922 \times 0.05}{1.846 \times 10^{-5}}$$

$$\Rightarrow R_e = 2497$$

Since air is in the turbulent flow regime then the Dittus and Boelter equation was applicable:

$$N_u = KR_e^{0.6}P_r^{0.33}$$

$$N_u = 0.32 \times (2497)^{0.6} \times 0.707^{0.33}$$

$$\Rightarrow N_u = 31.08$$

Therefore convection heat transfer coefficient was given by:

$$h_a = \frac{N_u \times k_a}{d_o} = \frac{31.18 \times 0.02624}{0.05}$$

$$\Rightarrow h_a = 16.36\ W/m^2K$$

Therefore overall heat transfer coefficient was given by:

$$\frac{1}{U} = \frac{1}{287.8} + \frac{0.002}{204} + \frac{1}{16.36}$$

$$\Rightarrow U = 15.5\ W/m^2K$$

Log mean temperature difference, ΔT_m was expressed as:

$$\Delta T_m = \frac{(T_{f2} - T_{a2}) - (T_{f1} - T_{a1})}{In\left(\frac{T_{f2} - T_{a2}}{T_{f1} - T_{a1}}\right)} = \frac{(300-70)-(600-26)}{In\left(\frac{300-70}{600-26}\right)}$$

$$\Rightarrow \Delta T_m = 376\,K$$

Therefore heat transfer area was calculated as:

$$A_s = \frac{Q_L}{U\Delta T_m}\text{, but } Q_L = m_a C_{pa}(T_{a2} - T_{a1}) = 0.27 \times 1.005 \times 44 = 12\ kW$$

$$A_s = \frac{12 \times 1000}{15.5 \times 376}$$

$$\Rightarrow A_s = 2.06\ m^2$$

Number of tubes

Let N_t = number of tubes

We know that heat transfer surface area is given by: $A_s = \pi d_o N_t L$

Where: d_o = external diameter of tube

L = length of the tube which is equal to the width of the dryer given as, *1.22m*

$$N_t = \frac{A_s}{\pi \times d_o \times L} = \frac{2.06}{\pi \times 0.05 \times 1.22}$$

$$\Rightarrow N_t = 11$$

Width of the heat exchanger shell

Let W_h width of heat exchanger

$$W_h = 6 \times P_T + d_o + a$$

Where: a = allowance on both side taken as 150 mm

$$W_h = 6 \times 75 + 50 + 150$$
$$\Rightarrow W_h = 650\, mm$$

Chimney draught

Known parameters

1) Ambient temperature, $T_{a1} = 26^o C$

2) Flue gas outlet temperature, $T_{f2} = 300^o C$

3) The height of the chimney was taken as, $z = 2\ m$

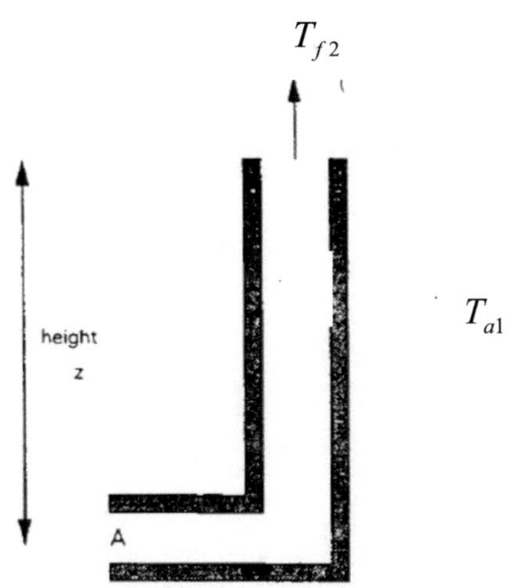

Figure 2: Chimney buoyancy forces.

Figure 2 above shows the pressure difference due to the buoyancy of the column of hot flue gas in the chimney which is simply equal to the relative weights of the columns of hot gas and surrounding air.
Flue gas draught can be calculated using the following expression;

$$\frac{\Delta p}{z} = 3{,}429\left[\frac{1}{T_{f2}} - \frac{1}{T_{a1}}\right] = 3{,}429\left[\frac{1}{573} - \frac{1}{299}\right]$$
$$\frac{\Delta p}{2} = -5.484$$
$$\Rightarrow \Delta p = -10.97\ Pa$$

Biomass stove design

The biomass stove main dimensions were designed by considering:

The thermal energy (Q_L) required to supply hot drying air at 70^{0}C which was found to be 12 kW.

The calorific value of the briquettes (CV_b) which ranges between 16 MJ/kg to 20 MJ/kg. An average value of 18 MJ/kg was taken.

Assuming that efficiency of heat generation (η_{st}) of the stove to be 86%.

Average density of briquette (ρ_b) was assumed to be 660 kg/m^3.

Refractory lining for the stove, $R_l = 40\ mm$

Fuel consumption rate

Fuel consumption rate (FCR) was given by:

$$FCR = \frac{Q_L}{CV_b \times \eta_{st}} = \frac{12 \times 3600}{18000 \times 0.86}$$
$$\Rightarrow FCR = 2.8\ kg/h$$

The biomass stove was designed to operate continuously for two hours before the next loading. Therefore the fuel weight per load was given as:

$$W_{load} = FCR \times Time = 2.8 \times 2$$
$$\Rightarrow W_{load} = 5.6\ kg$$

RESULTS

Main solar tunnel dryer construction details

The dryer consisted of a flat plate air heating collector and a drying tunnel fabricated as one unit. The top surfaces of the collector and drying chamber were designed in triangular shape with an inclination of 24^0 on both sides in order to increase the area of irradiance. Both the collector and the drying units were covered with UV-stabilized plastic sheet. Black paint was used as an absorber in the flat plate collector to supply the required hot air into the drying chamber during normal daytime operation of the dryer when sufficient solar radiation is available. The dryer was 1.24 m wide with a collector length of 3 m and a dryer length of 3 m. The main tunnel dryer frame was made of mild steel square hollow tube and angle lines with pieces of hardwood arranged at the bottom to act as insulation material to reduce heat loss from the bottom of the dryer. The pieces of hardwood at the bottom were covered with aluminium painted galvanized iron sheet to reflect energy incident on the surface. Similarly, the side walls of the tunnel chamber were made of aluminium coated galvanized sheet riveted on the main frame. (Figure 3)

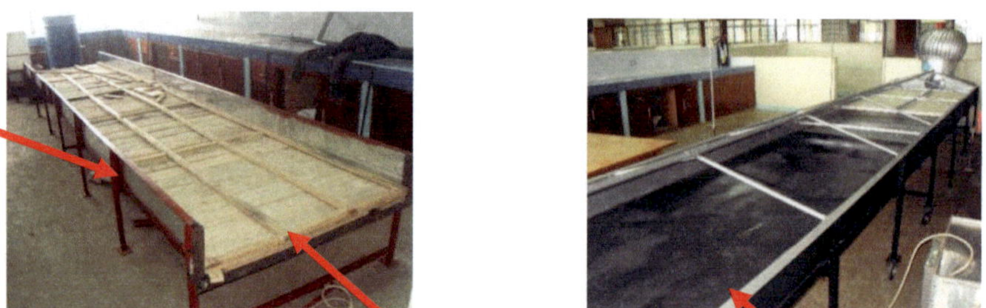

Main frame

Hardwood arrangement at the bottom of dryer

Black painted flat plate solar collector

Figure 3. Material arrangement and design for solar tunnel dryer.

The drying unit consisted of six trays made of perforated plastic mesh fixed on aluminium angle line frame by pop rivets. The trays, arranged horizontally, were fixed at a higher level of about 25 mm than the absorber plate to allow smooth flow for heated air on all sides of the dried product preventing the need to turn it during drying. Each tray had a loading area of 0.61 m^2 (Figure 4).

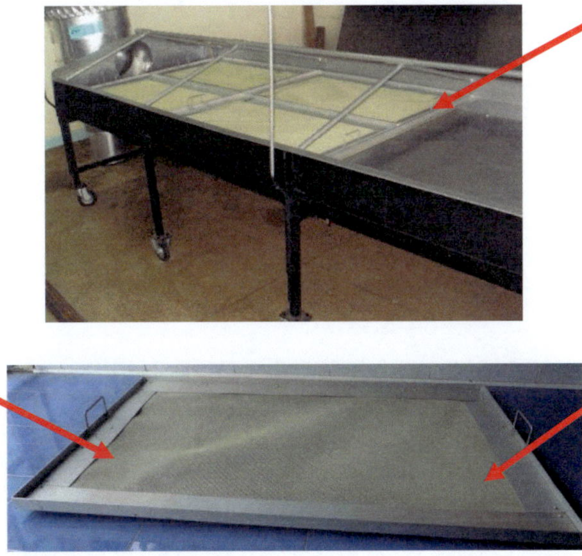

Figure 4. Tray details and arrangement on the dryer.

The dryer was designed in such a way that the UV-stabilized transparent cover could be lifted away from the trays during loading and unloading of the products (Figure 5).

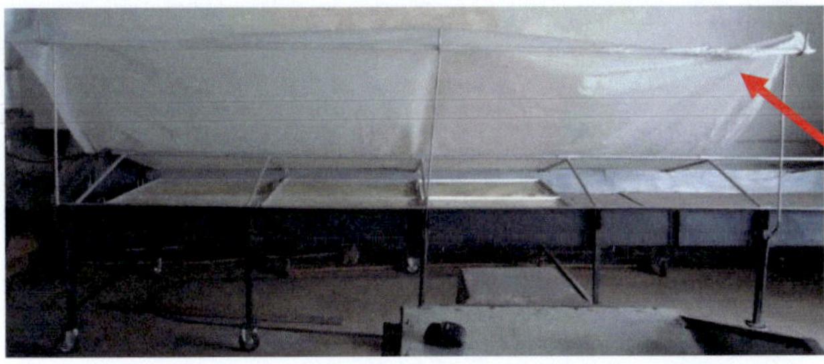

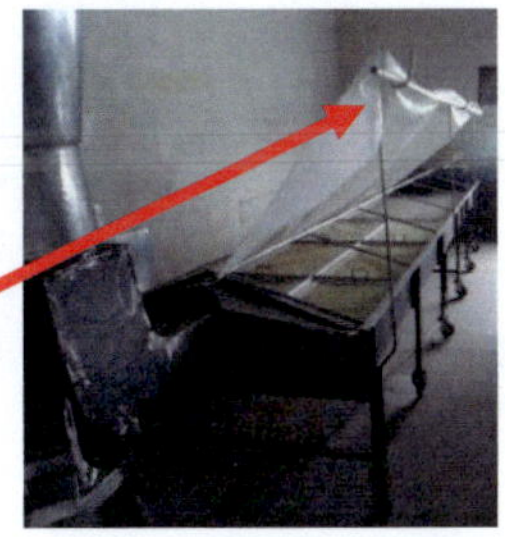

Figure 5. Illustration of how the UV sheet is lifted up during tray loading and unloading.

The tunnel dryer was fitted with a turbo-ventilator to provide the required air flow over the products to be dried. The turbo-ventilator depends entirely on the wind effect and exhaust air from the drying tunnel to operate. The centrifugal force caused by the spinning vanes creates a region of low pressure area which draws and throws out hot air from below as fresh cool air from outside comes in through the air inlet door located at the other end of dryer. The turbo-ventilator was placed on top of a stack fitted at the end of the drying chamber. The stack was made with an appropriate height of 3 m to achieve chimney effect and catch more wind for its operation. The turbo-ventilator rotor was mounted on bearing suspension with low friction such that as the rotor spins in the wind it expels air from the ventilator stack, hence air flows in the tunnel dryer (Figure 6). The whole dryer system was designed to be dissembled during transportation and was fitted with caster wheels to make it portable.

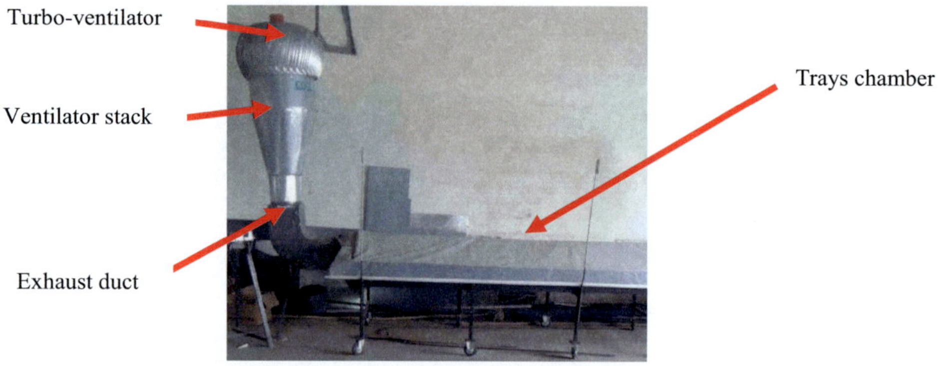

Figure 6. Turbo-ventilator on the dryer.

Solar-biomass heat exchanger system construction details

The biomass stove-heat exchanger system was designed to complement the solar operation of the dryer, and sustain the drying process even during cloudy weather. However, it can also be used to extend the period of drying beyond sunshine hours. Note that channel stove of rectangular shape and cross flow type heat exchanger were chosen. A biomass stove-heat exchanger assembly was placed at the inlet of the tunnel dryer to improve the reliability of the dryer and the quality of the dried product. The heat exchanger was of cross-flow type with flue gas from the stove passing through the tubes and ambient air flowing over the tubes before entering the dryer. A turbo-ventilator was used to force ambient air into the dryer, through the bank of heat exchanger tubes. The 'tubes' of the heat exchanger were connected to a biomass stove at one end and a chimney at the other through rectangular ducts. Eleven aluminium pipes with outer diameter 50 mm and inner diameter 44 mm were used to provide the required heat transfer area of the heat exchanger. The tubes were arranged in a staggered manner (Figure 7) to increase turbulence. The tube bank was placed inside mild steel shell of length: 1.72 m, width: 0.6 m, and height: 0.16 m.

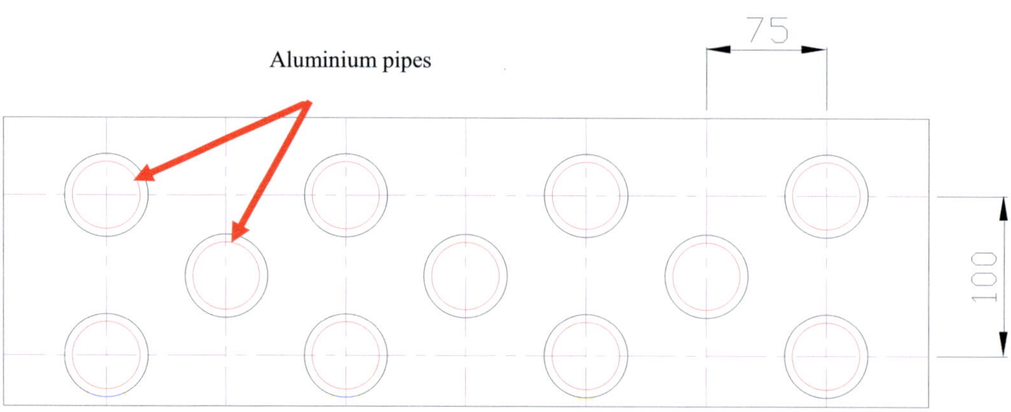

Figure 7. Arrangement of tubes in the heat exchanger (triangular type).

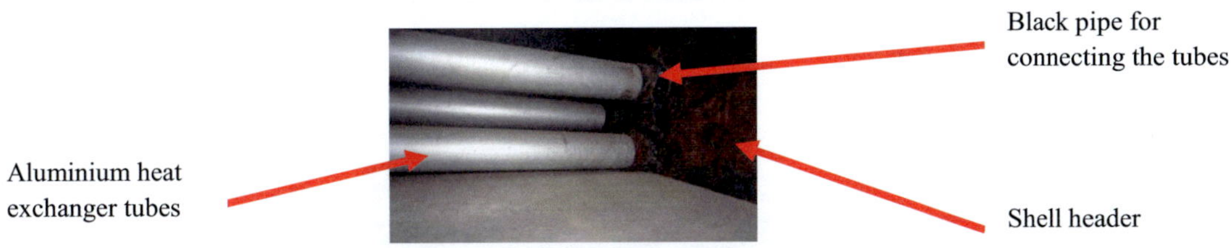

Figure 8. Aluminium pipes for the heat exchanger attached to the shell header.

Figure 8 above illustrates how the heat exchanger aluminium tubes were attached to the shell header. The tubes were fitted inside black pipe welded on the header from the inside leaving allowance for expansion. The outside of the header was insulated with fire bricks joined together using castable refractory cement (Figure 9). The header was made of a mild steel plate of 3 mm thickness.

Figure 9. Refractory lining on the header.

One side of the heat exchanger was provided with three air inlet doors with plates which can be adjusted by sliding up to partially open the door to control the amount of air entering the dryer, thus determining the outlet temperature of the heat exchanger. The doors were also fitted with wire mesh to prevent insects and flies from entering the dryer (Figure 10).

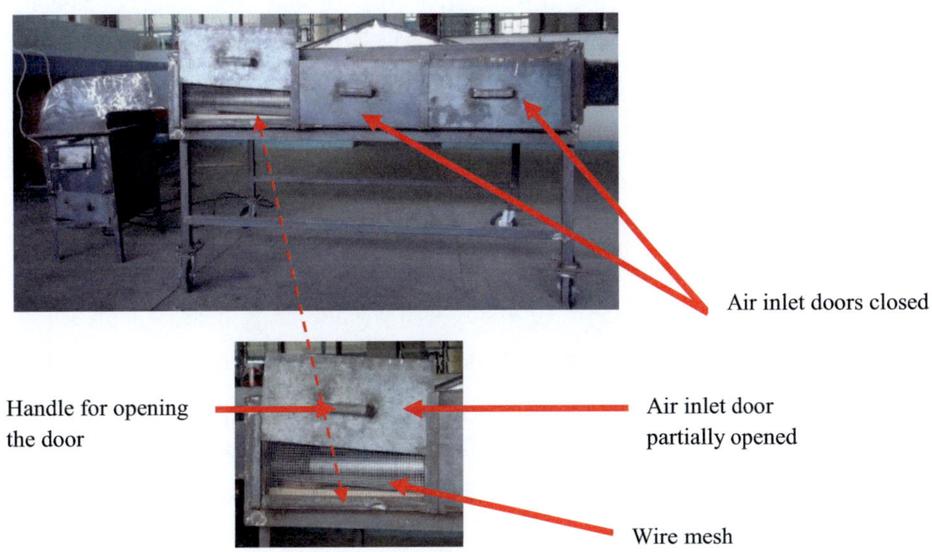

Figure 10. Air inlet door design with wire mesh

The other side of the heat exchanger was made open to discharge heated air into the drying chamber (Figure 11).

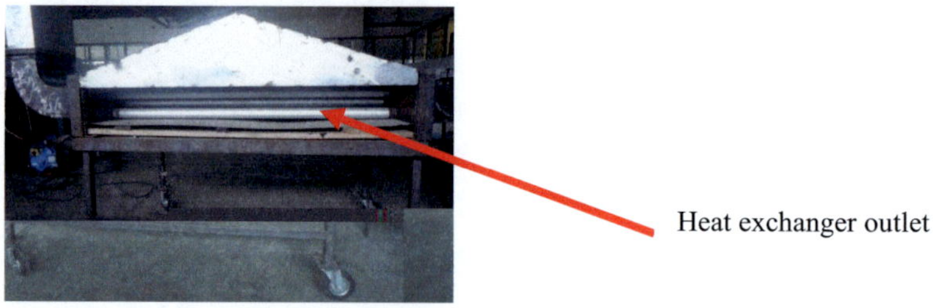

Figure 11. Outlet of the heat exchanger.

The main frame of the stove was made of mild steel angle line of 25 mm x 25 mm and mild steel sheet of 2 mm thickness. The inside walls of the stove was insulated with a 30 mm thick refractory lining using castable refractory cement. The stove was designed with two doors one for loading the fuel and the other for removing ash from the stove as well as controlling the primary air for combustion (Figure 12).

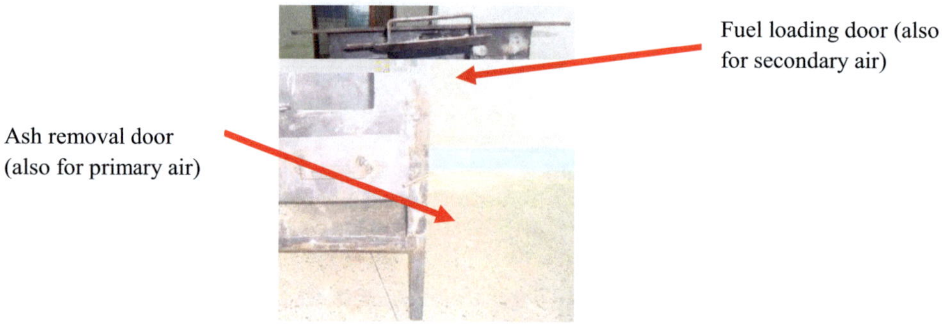

Figure 12. Two doors on the stove.

The stove had a grate made of mild steel round bars of 15 mm diameter designed to allow ash to fall through it into the ash pit and also to increase the efficiency and quality of combustion (Figure 13).

Figure 13. Grate design.

An ash scraper (Figure 14) was fixed below the grate to remove ash that accumulates inside the reaction chamber. The ash scraper is especially useful while using fuels of high ash content, such as rice husk briquettes, since accumulated ash could otherwise block the flow of fresh fuel from the fuel chamber into the reaction chamber. The ash scraper was designed to slide through a cylindrical guide bush welded to the body of the stove (Figure 14). For easy assembling, the slider rod was connected to the 'fingers' of the scraper by a threaded joint. The ash scraper is operated by sliding it in and out horizontally, generally once in 10-20 minutes depending on the ash content of the fuel.

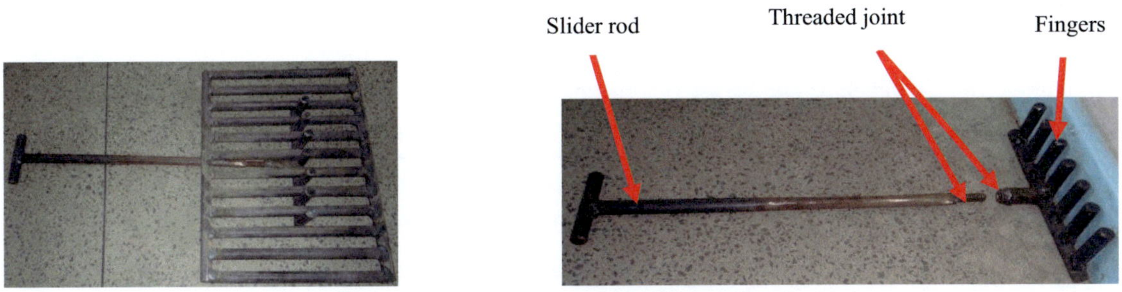

Figure 14. Ash scraper design and its assembly below the grate.

Figure 15. Position of the ash scraper and the grate in the stove.

Figure 15 illustrates the assembly of the ash scraper and the grate in the stove before insulating with castable refractory cement.

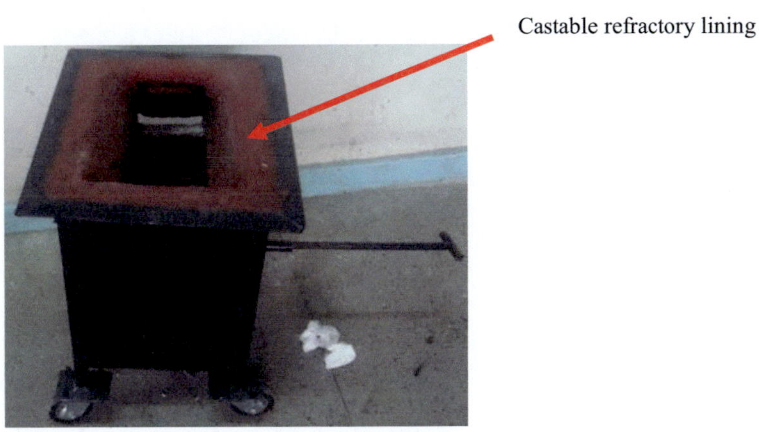

Figure 16. Complete assembled biomass stove.

Figure 16 shows the complete assembled biomass stove insulated with castable refractory lining. Note that fuel (briquette pieces of 40 x 40 mm) is loaded through the side doors up to the top level of the stove and then ignited using wood chips and kerosene. Primary air for combustion is taken through the ash pit door, which is kept open during operation. Secondary air is taken through the fuel door, which also is kept open partially. The stove has a high chimney to produce hot flue gas in natural draught. The chimney was designed for a flue gas flow rate of about 139 m^3/hour. A rectangular shape was chosen for the chimney, which provides the draft needed for flue gas from the stove to flow through the heat exchanger tubes, has a cross-sectional area of 187 mm x 300 mm and height 2,000 mm. The biomass stove, heat exchanger and chimney were connected together using rectangular ducts with asbestos gasket (Figure 17). Figure 18 illustrates the orthographic view of the assembled solar biomass hybrid tunnel dryer.

Figure 17. Complete assembled biomass stove heat exchanger system.

CONCLUSION

It can be concluded that the design and fabrication of the solar biomass hybrid tunnel dryer was successful. Future work will focus on performance evaluation of the dryer.

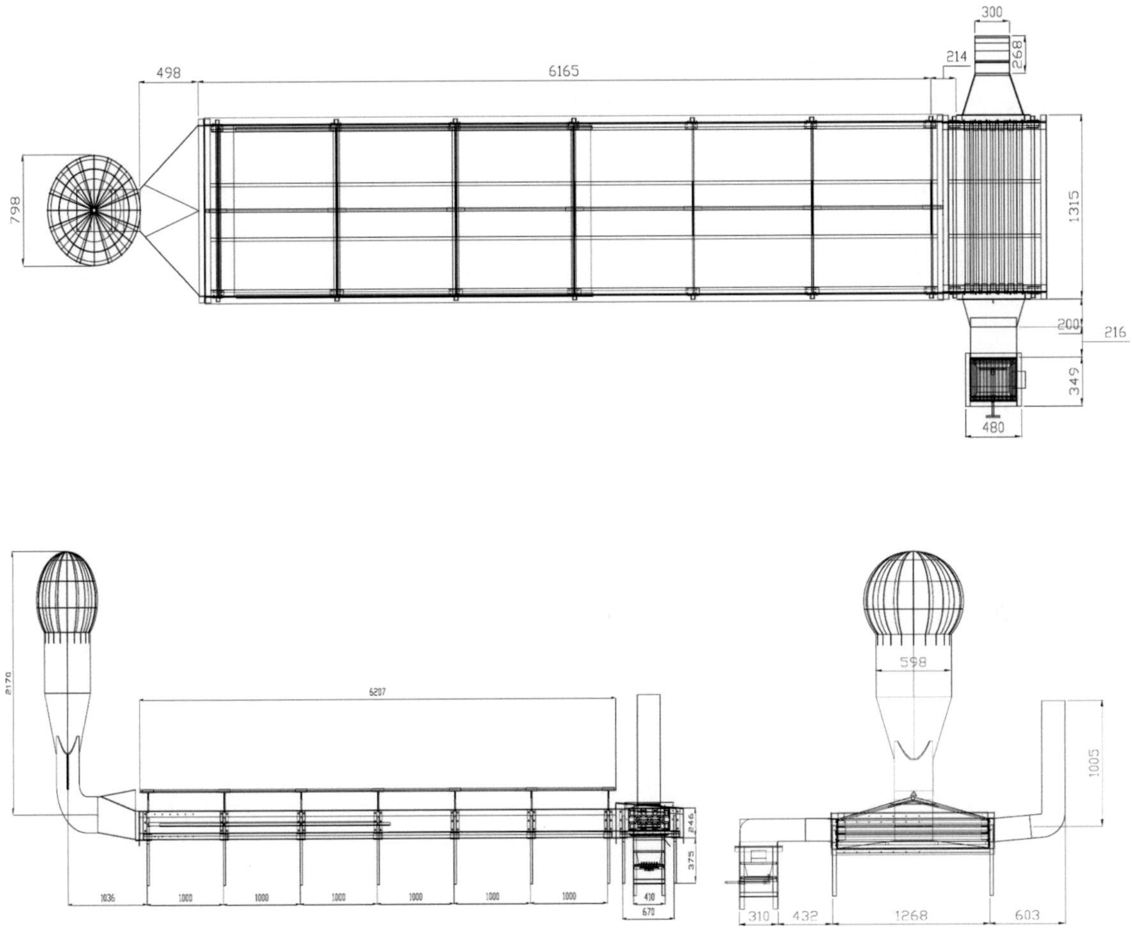

Figure 18. Orthographic drawings of solar-biomass hybrid tunnel dryer.

12. Effect of water hyacinth (*Eichhornia crassipes* [mart.] solms) compost on growth and yield parameters of maize (*Zea mays*)

Newton Osoro[1]*, John M. Maingi[1], Omwoyo Ombori[1], John O. Muoma[2], Alice Amoding[3], Dative Mukaminega[4], Fanuel Kawaka[5] and Morris Muthini[6]
[1]Department of Plant and Microbial Sciences, Kenyatta University, Kenya.
[2]Department of Biological Sciences, Masinde Muliro University of Science and Technology, Kenya.
[3]Department of Soil Science, Makerere University, Uganda.
[4]Faculty of Applied Sciences, Kigali Institute of Science and Technology, Rwanda.
[5]Department of Pure and Applied Sciences, Technical University of Mombasa, Kenya.
[6]Department of Biotechnology, Kenyatta University, Kenya.
**Corresponding author: tony_soro@yahoo.com*

SUMMARY

Water hyacinth (*Eichhornia crassipes*) has become a problem in water bodies all over the world. Various methods used to eradicate the macrophyte have yielded little success. The weed has high content of the nutrients absorbed from its environment. Hence there is need to assess the potential of utilizing it for commercial purposes such as compost to substitute the use of inorganic fertilizers which contributes to climate change. The objective of the study was to evaluate the effects of water hyacinth compost prepared with various cultures on growth and yield parameters of maize. Field experiments were carried out at Otonglo Division in Kisumu county and greenhouse trials were carried out at Kenyatta University to evaluate the effects of water hyacinth compost prepared with various cultures on growth and yield parameters of maize. The water hyacinth compost was prepared using effective microorganisms (EM) solution, cow manure and molasses separately as starter cultures for composting. The effect of compost on crop production was assessed by applying the compost to maize (H513) on separate growth pots in the greenhouse and plots at the field respectively. The treatments applied included compost prepared with EM (8.837 g/pot), compost prepared with cow manure (8.837 g/pot), compost prepared with molasses (8.837 g/pot), diammonium phosphate (DAP) fertilizer (70.7 mg nitrogen/pot) and control (without fertilizer). For the field trials, water hyacinth compost was applied at a rate of 5,000 kg/hectare and DAP at 40 kg nitrogen/hectare. The experiments were laid out in a randomised complete block design with five treatments and three replications for each treatment in the greenhouse and four treatments replicated four times in two farmers' fields. Results showed that application of water hyacinth compost and DAP significantly influenced positively ($P \leq 0.05$) the growth attributes of maize evaluated including plant height, shoot dry weight, and root dry weight, number of leaves per plant and root collar diameter. Yield parameters such as 100 seed-weight, grain yield and stover weight were not significantly ($P > 0.05$) influenced by various treatments. It can therefore be concluded that water hyacinth which is locally available and in large quantities (especially in lake Victoria) can be composted to prepare organic fertilizers and effectively used as an organic soil amendment to increase maize production.

INTRODUCTION

Recent census report on the population of the Lake Victoria Basin shows that about 80% of the population is engaged in agricultural production (Albinus et al., 2008). These farmers continuously cultivate their land leading to a reduction in soil fertility and reduced crop productivity (Ojiem et al., 2000). Chemical fertilizers which are used by farmers in the region are expensive; cause eutrophication hence algal blooms emerge and increase the overall greenhouse gases such as the oxides of carbon and nitrogen leading to environmental destruction (Khanal, 2009). The economic and environmental costs as a result of the use of the chemical fertilizers together with their negative impacts to the environment are now of global concern hence there is need to shift to sustainable organic farming practices. The use of organic fertilizers from unexploited natural resources perceived as weeds, such as water hyacinth, would be a better alternative to improve soil fertility and increase crop yield.

Water hyacinth (*Eichhornia crassipes* [Mart.] *Solms*) is a free floating macrophyte (in the family Pontedericeae, Liliales) that grows in fresh water, but may be rooted in the mud, from where it draws its nutrients. Wilson et al. (2005) recognized water hyacinth as the most harmful aquatic weed in the world due to its negative effects on people's livelihoods and waterways. Efforts to control the weed through chemical, physical and biological methods have all failed. However, significant economic benefits can be be obtained from water hyacinth (Abdelsabour, 2010; Amoding et al., 1999; Anjanabha and Kumar, 2010). The weed is a valuable source of macronutrients such as phosphorus, nitrogen and potassium that are essential for plant nutrition (Center et al., 2002; Sahu et al., 2002; Woomer et al., 2000). Due to the availability of water hyacinth in Lake Victoria in large quantities; it can be used to produce organic fertilizer. The compost could then be applied to plants as a source of nitrogen, phosphorus and potassium (NPK) which are the macro nutrients that limit crop growth (Kwabiah et al., 2003; Wasonga et al., 2008).

Maize (*Zea mays*) is the most important staple food in Kenya providing about 40% of the populations' caloric requirements and forming the major component of meals. Kenya produces an average of 30 million bags (1 bag – 90 kg) of maize annually most of which is from smallholder farms with an average of 1.5-2.5 acres per household (Tungani et al., 2003). At smallholder farmer level, yields are low averaging 1.5 tons/hectare compared to research centre yields of 6-8 tons/hectare (Jaetzold, 2006). The yield of maize could improve as a result of improved soil fertility management through compost application (Achieng et al., 2001).The objective of the study was therefore to assess the effectiveness of the organic fertilizer from composted water hyacinth on growth and yield of maize.

MATERIALS AND METHODS

The field experiments were conducted at Korando sub-location, Otonglo division, Kisumu County in Lake Victoria Basin from November 2011 to August 2012. The site location is latitude 0° 20' S and 0° 50' S and longitude 33° 20' E and 33° 50' E at an elevation of 1,131 m a.s.l. Planting was done between April and July. Water hyacinth was harvested manually, sun-dried and chopped into small pieces of about 5 cm. Above ground closed aerobic heap design was used to prepare the compost. The water hyacinth compost was prepared using effective microorganisms (EM) solution, cow manure and molasses separately as starter cultures for composting. Clean plastic pots (15 cm diameter x 20 cm height) were used. The pots were filled with soil and moistened by applying water until the soil was thoroughly wet. Various fertilizer treatments were applied to each soil per pot and thoroughly mixed. Five viable maize seeds per pot of uniform size and shape were sown at a depth of 3 cm and one week after emergence, they were thinned to three seedlings per pot. The treatments administered were as follows: maize and water hyacinth compost with EM (WHE), maize and water hyacinth compost with cattle manure (WHC), maize and water hyacinth compost with molasses (WHM), maize and nitrogen fertilizer (DAP at 70.7 mg nitrogen/pot), maize with no fertilizer (control). Water hyacinth compost was applied at 8.837 g/pot. Watering of the pots was carried out by adding 100 ml water on alternate days to each pot. Each treatment was replicated thrice on separate blocks and the pots arranged in a randomized complete block design. Sampling of plants was done at 15, 30 and 45 days after emergence (DAE). The number of leaves, root collar diameter and height of the maize plant sampled was measured. The shoots and roots of the plants were separated, oven dried at 60°C to constant weight and their dry weights measured. Each greenhouse experimental set up was stopped after 45 days. The experiment was repeated three times.

At the field, two researcher managed farmers' fields (Chief's farm and Nyaguti's farm) were prepared by disc ploughing followed by harrowing. Each field was demarcated into four main blocks with four plots separated by 1 m paths to minimise inter-plot interference. Viable and uniform seeds for maize were planted in the plots in the field in April, 2012. Two seeds per hole were planted with various treatments and two weeks after germination, they were thinned to one plant per hole after the first weeding. An inter-row spacing of 0.75 m and intra-row spacing of 0.3 m was used. Twenty eight maize plants per plot were retained after thinning. First weeding was carried out in both farms at two weeks after emergence, after which the plants were top dressed using CAN fertilizer (CAN at 40 kg/hectare) for those plots that had been planted with nitrogen fertilizer. Second weeding was carried out four weeks later. The layout in each field was a randomised complete block design with four blocks. The treatments administered were the same as the greenhouse excluding WHM. Water hyacinth compost was applied at 5000 kg/hectare while DAP (NPK ratio was 18:46:0) was applied at 40 kgN/hectare. CAN (26% nitrogen) was applied at 40 kg nitrogen/hectare to those plots that had been treated with DAP fertilizer. The data was analysed using one way analysis of variance with SPSS computer software version 11.5 for Windows. Means were separated using Tukey's honest significant difference at 5% level.

RESULTS AND DISCUSSION

The mean shoot dry weight per maize plant was significantly different at 30 DAE (P = 0.000) and 45 DAE (P = 0.000) among various treatments but not at 15 DAE (P = 0.565). DAP treatment produced the highest mean shoot dry weight while control produced the least at all days of sampling (Table 1). Compost and DAP increased the biomass of the shoot more compared to the control probably due to the release of considerable amounts of nutrients, especially nitrogen and phosphorus, to the plant during mineralization that were utilized in chlorophyll formation for photosynthesis hence better plant development (Kamanu et al., 2012). The shoot dry weight recorded for plants grown on soil amended with DAP was significantly higher in maize compared to other treatments. This could be attributed to the availability of nitrogen and phosphorous more readily in inorganic fertilizer compared to compost that releases its nutrients slowly. A significant increase in the shoot dry weight of plants with application of DAP has also been reported (Cheminingwa et al., 2007; Otieno et al., 2009; Kamanu et al., 2012).

The mean root dry weight per plant of maize grown under different soil amendments were significantly different on the 30th and 45th DAE (P = 0.000 at 30 DAE and P = 0.000 at 45 DAE). The control gave the least mean root dry weight during the three stages of sampling (Table 2). The highest mean root dry weight, though statistically similar to other treatments on the 15th DAE but statistically different from other treatments on the 30th DAE, was recorded among plants treated with WHM while DAP treatment yielded the highest mean root dry weight (3.50 g) on the 45th DAE which was significantly different from all other treatments (Table 2). This could have been attributed to the high amounts of organic matter in compost which could have increased the moisture retention of soil, improved dissolution of nutrients particularly phosphorus and soil structure hence better root growth and nutrient uptake (Olupot et al., 2004). These results conform to findings of Amitava et al. (2008) and Chukwuka and Amotayo (2008, 2009).

The least mean plant height in the greenhouse was recorded among maize plants under the control experiment (13.42 cm) while the highest mean height was noted when the soil was amended with DAP (36.87 cm) throughout the experiment (Table 3). This could have been due to the availability of nutrients from chemical fertilizers that offer readily soluble nutrients which are instantly availed to the plant unlike compost that provides the nutrients slowly but steadily. The physical properties of the soil could also have improved, thus better uptake of nutrients by the crop hence better crop development (Amitava et al., 2008). Similar findings have been reported by Achieng et al. (2010); Mohsin et al. (2012); Shah et al. (2009) and Yagoub et al. (2012).

The root collar diameter among maize plants was also influenced by various soil amendments. Highly significant root collar diameter measurements were recorded on the 15th DAE (P = 0.002), 30th DAE (P = 0.000) and 45th DAE (P = 0.000). DAP treatment produced plants with the highest root collar diameter (11.02 mm) on the 45th DAE and the 30th DAE (10.42 mm) while WHM produced plants with the highest root collar diameter (7.51 mm) on the 15th DAE (Table 4). The control experiment led to the least significant collar diameter among all treatments in all sampling days (Table 4). The availability of nutrients in DAP and compost were utilized by the plants to amass their dry weights hence producing bigger plants with a statistically higher root collar diameter. Similar findings were reported by Chukwuka and Amotayo (2008, 2009).

Significant differences were observed in the number of leaves per maize plant grown under different soil amendments in the greenhouse on the 30th and 45th DAE (Table 5). The control produced the least number of leaves per plant all through although they were not significantly different from those produced by maize plants grown in the soil amended with other treatments ($P = 0.061$) on the 15th DAE. However, it was significantly different on the 30th ($P = 0.009$) and 45th ($P = 0.033$) DAE. The highest number of leaves of leaves was recorded when the soil was amended with DAP (8) while the control produced the least number (4) (Table 5). Adequate amounts of nitrogen supplied by either DAP or compost could have increased the number of leaves thus result to higher yields. Sanni and Adesina (2012) reported non-significant results on the number of leaves but a higher number of leaves on plants subjected to water hyacinth compost than the control at one week after application and significant results at three weeks after application.

Maize yield

Various soil amendments had no significant effect on the maize grain yield at both Nyaguti's farm ($P = 0.61$) and Chief's farm ($P = 0.148$). WHE treatment produced the highest amount while DAP produced the least yield in Nyaguti's farm. In Chief's farm, soil amendment with DAP resulted into the highest grain yield per hactare while WHE recorded the lowest yield (Table 6). The increased grain yield in the application of DAP could be attributed to the readily available nitrogen and phosphorus supplied by the fertilizers applied especially during grain filling and development period of the plants. Compost did not significantly increase grain yield due to its slow release of nutrients. Okoko and Makworo (2000) reported that continuous application of compost over time produced significantly better results than inorganic fertilizer due to its slow but steady release of nutrients but not in the first season of application. Manure effectively regulated the soil's acidity probably by binding the exchangeable aluminium ions in the acidic soils besides releasing plant nutrients slowly hence equally produced higher grain yields (Tejada et al., 2006; Onwonga and Lelei, 2010). The lower yields observed in the control could have been due to reduced nitrification rates and phosphorus fixation in the acidic soils that rendered nitrogen and phosphorous unavailable for plant uptake thus lower yield (Onwonga and Lelei, 2010). Chief's farm recorded lower grain yields than Nyaguti's farm probably due to lack of sufficient nitrogen in the soil which stressed the plant especially at the initial growth stages thus the crops exhibited chlorosis and retarded growth hence lower yields. The acidity of the soil could also have caused phosphorus fixation which made it unavailable to the plants hence no significant difference between the treated plots and control (Fardeau and Zapata, 2002; Kamanu et al., 2012).

Table 1. ^{X}Mean shoot dry weight (g) of maize grown in the greenhouse under different soil amendments.

Treatment	15 DAE	30 DAE	45 DAE
WHC	0.73±0.13^{a}	2.16±0.21^{ab}	4.24±0.41^{a}
WHE	0.68±0.09^{a}	2.4±0.19^{bc}	5.72±0.59^{a}
WHM	0.79±0.09^{a}	2.56±0.15^{bc}	4.01±0.1^{a}
DAP	0.79±0.18^{a}	3.06±0.29^{c}	11.07±2.11^{b}
Control	0.54±0.08^{a}	1.48±0.13^{a}	2.19±0.28^{a}

Table 2. ^{X}Mean root dry weight (g) of maize grown in the greenhouse under different soil amendments.

Treatment	15 DAE	30 DAE	45 DAE
WHC	0.15±0.04^{a}	0.77±0.08^{bc}	2.13±0.22^{a}
WHE	0.17±0.04^{a}	0.70±0.02^{a}	2.06±0.30^{a}
WHM	0.18±0.04^{a}	0.93±0.08^{c}	1.85±0.08^{a}
DAP	0.16±0.05^{a}	0.55±0.03^{ab}	3.50±0.53^{b}
Control	0.14±0.04^{a}	0.42±0.04^{a}	1.05±0.09^{a}

Table 3: ^{X}Mean plant height (cm) of maize grown in the greenhouse under different soil amendments.

Treatment	15 DAE	30 DAE	45 DAE
WHC	15.57±0.56^{a}	22.08±0.97^{ab}	26.17±1.19^{a}
WHE	16.32±0.47^{a}	23.58±0.72^{ab}	32.98±2.05^{bc}
WHM	15.50±1.15^{a}	23.82±1.53^{ab}	28.87±1.69^{ab}
DAP	16.95±1.26^{a}	26.58±2.09^{b}	36.87±1.96^{c}
Contrrol	13.42±0.43^{a}	20.32±1.27^{a}	23.67±0.79^{a}

Table 4. Root collar diameter (mm) of maize grown in the greenhouse under different soil amendments.

Treatment	15 DAE	30 DAE	45 DAE
WHC	7.50±0.18^{b}	9.33±0.48^{b}	7.28±0.21^{ab}
WHE	6.67±0.21^{ab}	9.50±0.48^{b}	9.23±0.72^{bc}
WHM	7.51±0.18^{b}	9.58±0.20^{b}	7.67±0.49^{ab}
DAP	7.25±0.66^{b}	10.42±0.30^{b}	11.02±0.57^{c}
Control	5.50±0.22^{a}	6.92±0.33^{a}	6.00±0.2^{a}

Table 5. Number of leaves per plant of maize grown in the greenhouse under different soil amendments.

Treatment	15 DAE	30 DAE	45 DAE
WHC	5±0.31^{a}	6±0.37^{ab}	7±0.21^{ab}
WHE	5±0.17^{a}	6±0.26^{ab}	7±0.33^{ab}
WHM	5±0.21^{a}	7±0.21^{b}	7±0,40^{ab}
DAP	5±0.31^{a}	7±0.33^{b}	8±0.22^{b}
Control	4±0.21^{a}	5±0.31^{a}	6±0.17^{a}

Key to Table 1, 2, 3, 4 and 5

^{X}Mean±SE.

DAE - Days after emergence.

Means within the column followed by the same letters are not significantly different at P>0.05.

WHC - Water hyacinth compost prepared using cattle manure treatment.

WHM - Water hyacinth compost prepared using molasses treatment.

WHE - Water hyacinth compost prepared using EM treatment.

DAP - Diammonium phosphate (Nitrogen fertilizer)

Control - No added fertilizer.

Table 6. Grain yield of maize grown in the two sites under different soil amendments.

Treatment	Nyaguti's farm	Chief's farm
WHE	8622.75±2739.1[a]	2991.16±194.43[a]
WHC	6541.28±738.44[a]	4440.9±812.45 [a]
DAP	6047.83±902.09[a]	4568.41±531.54[a]
CONTROL	6169.47±584.87[a]	3105.25±590.43[a]

[X]Mean ± SE.
DAE - Days after emergence.
Means (n=12) within the column followed by the same letters are not significantly different at P>0.05.
WHC - Water hyacinth compost prepared using cattle manure treatment.
WHE - Water hyacinth compost prepared using EM treatment.
DAP - Diammonium phosphate (nitrogen fertilizer).
Control - No added fertilizer.

Application of various treatments on soil did not significantly affect the dry weight of 100 maize grains in both farms (Table 7). DAP gave the highest dry weight while WHC gave the lowest in Nyaguti's farm but the difference was insignificant (P = 0.369). At Chief's farm WHC treatment gave the highest weight while control gave the least weight (P = 0.37). Statistically similar results in the weight of 100 maize seeds among various treatments have also been reported by Mburu et al. (2011) and Yagoub et al. (2012).

Table 7. 100 seed weight of maize grains grown in the two sites under different soil amendments.

Treatment	Nyaguti's farm	Chief's farm
WHE	44.48±1.26[a]	35.17±1.52[a]
WHC	41.73±0.5[a]	37.96±2.29[a]
DAP	43.78±1.72[a]	35.2±1.07[a]
CONTROL	42.22±0.99[a]	32.63±2.8[a]

[X]Mean ± SE.
DAE - Days after emergence.
Means (n=12) within the column followed by the same letters are not significantly different at P> 0.05.
WHC - Water hyacinth compost prepared using cattle manure treatment.
WHE - Water hyacinth compost prepared using EM treatment.
DAP - Diammonium phosphate (nitrogen fertilizer).
Control - No added fertilizer.

Table 8. Stover yield of maize grown in the two sites under different soil amendments.

Treatment	Nyaguti's farm	Chief's farm
WHE	20425.1±2784.1[a]	18612.22±485.34[a]
WHC	22450.0±2242.2[a]	15832.0±1535.2 [a]
DAP	25860.9±2419.7[a]	20524.9±6941.2[a]
CONTROL	20406.7±2590.3[a]	18930.0±2257.6[a]

[X]Mean ± SE.
DAE - Days after emergence.
Means (n=12) within the column followed by the same letters are not significantly different at P>0.05 using Tukey's honest significant difference at 95% confidence level.
WHC - Water hyacinth compost prepared using cattle manure treatment.
WHE - Water hyacinth compost prepared using EM treatment.
DAP - Diammonium phosphate (nitrogen fertilizer).
Control - no added fertilizer.

DAP amendment led to the heaviest stovers in the two farms but was not statistically different from other treatments ($P = 0.397$ in Nyaguti's farm; $P = 0.844$ in Chief's farm). Control treatment produced the lightest stovers in Nyaguti's farm while WHC produced the least weight in Chief's farm (Table 8). The stover weight after harvest was higher in plants grown in soil amended with DAP compared to other amendments probably due to release of considerable amounts of nutrients, especially nitrogen and phosphorus, to the plant during mineralization that were utilized for photosynthesis hence better plant development (Kamanu et al., 2012).

CONCLUSION

Water hyacinth compost increased the growth parameters of maize. This is therefore an indicator of better yields if proper farm management is practiced under favourable environmental conditions. DAP also performs well in crop production but its adverse effects on the soil and the environment makes its use undesirable. Although the field results indicate that maize production without fertilizer application in some farms is possible, there is fear that without replenishment, the macronutrients could be depleted leading to low yields. Therefore, compost from water hyacinth which is locally available, plentiful and cheap can be effectively used as an organic soil amendment for soil restoration and crop production.

ACKNOWLEDGEMENT
We thank VICRES (Lake Victoria Research Initiative) for funding the project.

REFERENCES

Abdel-sabour MF. (2010). Water hyacinth: available and renewable resource. Electronic Journal of Environmental Agriculture and Food Chemistry, 9(11): 1746-1759.

Achieng JO, Friesen D, Odongo O, Odendo M. (2001). Sustainability of fertilizer use in maize production in western Kenya through provision of credit. African Crop Science Conference Proceedings, 5:601-604.

Achieng JO, Ouma G, Odhiambo GO, Muyekho F. (2010). Effect of farmyard manure and inorganic fertilizers on maize production on Alfisols and Ultisols in Kakamega, western Kenya. Agriculture Biology Journal of Nature 1(4): 430-439.

Albinus MP, Obando J, Bamutaze Y. (2008). Effects of land use practices on livelihoods in the transboundary sub-catchments of the Lake Victoria Basin. African Journal of Environmental Science and Technology, 2(10):309–317.

Amitava R, Sarkar N, Debashish S. (2008). Influence of organic manures on productivity of two varieties of rice. Journal of Central European Agriculture, 9(4):629-634.

Amoding A, Muzira NR, Bekunda MA, Woomer PL. (1999).Bioproductivity and decomposition of water hyacinth in Uganda.African Crop Science Journal,7:433-439.

Anjanabha B, Kumar P. (2010).Water hyacinth as a potential biofuel crop. Electronic Journal of Environmental, Agriculture and Food Chemistry, 9(1):112-122.

Center TD, Hill MP, Cordo H, Julien MH. (2002). Water hyacinth.In Van Driesche, R (ed.). Biological control of invasive plants in Eastern U.S. USDA Forest Service Publication FHTET:41-64.

Chemining'wa GN, Muthomi JW, Theuri SW. (2007). Effect of Rhizobia inoculation and starter N on nodulation, shoot biomass and yield of grain legume. Asian Journal of Plant Sciences, 320(6):1113-1118.

Chukwuka KS, Omotayo OE. (2008). Effects of *Tithonia* Green Manure and Water Hyacinth Compost Application on Nutrient Depleted Soil in South-Western Nigeria. International Journal of Soil Science, 3(2):69-74.

Chukwuka KS, Omotayo OE. (2009). Soil fertility restoration potentials of *Tithonia* green manure and water hyacinth compost on a nutrient depleted soil in South Western Nigeria using *Zea mays* L. as test crop. European Journal of Soil Biology, 1:20-30.

Fardeau JC, Zapata F. (2002). Phosphorus fertility recapitalisation of nutrient depleted tropical acid soils with reactive phosphate rock. An assessment using the isotopic exchange technique. Nutrient Recycling in Ecosystem, 63:69-79.

Jaetzold R. (2006). Management handbook volume II (2^{nd}ed). Ministry of Agriculture, Livestock and Fisheries, Nairobi.
Kamanu JK, Chemining'wa GN, Nderitu JH, Ambuko J. (2012). Growth, yield and quality response of snap bean (*Phaseolus vulgaris L.*) plants to different inorganic fertilizers applications in central Kenya. Journal of Applied Biosciences, 55:3944–3952.
Khanal RC.(2009). Climate change and organic agriculture.Journal of Agriculture and Environment, 10:100-110.
Kwabiah AB, Stoskopf NC, Palm CA, Voroney RP, Rao MR, Gacheru E. (2003). Phosphorus availability and maize response to organic and inorganic fertilizer inputs in a short term study in Western Kenya. Agricultural Ecosystems and Environment 95:49-59.
Mohsin AU, Ahmad J, Ahmad AU, Ikram RM, Mubeen K. (2012). Effect of nitrogen application through different combinations of urea and farm yard manure on the performance of spring maize (*Zea mays* L.) The Journal of Animal and Plant Sciences, 22(1):95-198.
Ojiem JO, Mureithi JG, Okuosa EA. (2000). Integrated management of legume green manure, farmyard manure and inorganic nitrogen for soil fertility improvement in Western Kenya.Participatory technology development for management by smallholders in Kenya.KARI, Nairobi, Kenya. pp 97-102.
Okoko EN, Makworo S. (2000). Evaluation of the effect of compost and inorganic fertiliser on maize yield in Nyamira district, South West Kenya. KARI.
Olupot G, Etiang J, Ssali H, Nahasirye M. (2004). Sorghum yield response to Kraal manure combined with mineral fertilizers in Eastern Uganda. Muarik Bulletin, 7:30-37.
Onwonga RN, Lelei JJ.(2010). Mineral Nitrogen and microbial Biomass Dynamics under Different Acid Soil Management Practices for Maize Production. Journal of Agricultural Sciences, 2(1):16-30.
Otieno PE, Muthomi JW, Chemining'wa GN, Nderitu JH. (2009). Effect of rhizobia inoculation, farm yard manure and nitrogen fertilizer on nodulation and yield of food grain legumes. Journal of Biological Sciences, 9(4):326-33.
Sahu AK, Sahoo SK, Giri SS. (2002).Efficacy of water hyacinth compost in nursery pond for larval rearing of Indian major carp, *Labeo robitta.*Bioresource Technology, 85:309–311.
Sanni KO, Adesina JM. (2012). Response of water hyacinth manure on growth attributes and yield of *Celosia argentea* L (Lagos Spinach). Journal of Agricultural Technology, 8(3):1109-1118.
Shah STH, Zamir MSI, Waseem M, Ali A, Tahir M, Khalid WB. (2009). Growth and yield response of maize (*Zea mays* L.) to organic and inorganic sources of nitrogen. Pakistan Journal of Life Social Sciences, 7(2):108-111.
Tejada M, Hernandez MT, Garcıa C. (2006). Application of two organic amendments on soil restoration: effects on the soil biological properties. Journal of Environment Quaity, 35:1010–1017.
Tungani JO, Woomer PL,Mukhwana E.J. (2003). Strategies of applying mineral fertilizers to an innovative maize-legume intercrop in western Kenya.African Crop Science Conference Proceedings, African Crop Science Society.
Wasonga CJ, Sigunga DO, Musandu AO. (2008). Phosphorus requirements by maize varieties in different soil types of Western Kenya.African Crop Science Journal, 16(2):161-173.
Wilson JR, Holst N, Rees M. (2005). Determinants and patterns of population growth in water hyacinth. Aquatic Botany, 81:51-67.
Woomer PL, Muzira R, Bwamiki D, Metutikka D, Amoding A, Bekunda MA. (2000). Biological management of water hyacinth waste in Uganda. Biology Agriculture Horticulture, 17:181–196.
Yagoub SO, Wigdan MA, Mariod AA. (2012). Effect of urea, NPK and compost on growth and yield of soybean *(Glycine max L.)*, in Semi-Arid Region of Sudan. International Scholarly Research Network Journal of Agronomy, 1:38-44.

13. Toxigenic potential of *Aspergillus* species occurring on maize kernels from Makueni and Nandi counties

Vincent Ayugi[1]*, Beatrice M. Nyongesa[1], Sheila Okoth[1], James Ochanda[1]
[1]School of Biological Sciences, University of Nairobi, Nairobi.
**Corresponding author: vayugi811@gmail.com*

SUMMARY

Two agro-ecological zones in Kenya were selected to compare the distribution in maize of *Aspergillus* spp. and their toxigenicity. These were Nandi county, which is the main maize growing region in the country but where no human aflatoxicoses have been reported, and Makueni county where most of the aflatoxicosis cases have occurred. 250 households were sampled in Nandi and 258 in Makueni, and *Aspergillus* was isolated from maize. *Aspergillus flavus* and *A. parasiticus* isolates were tested for the presence of *aflD* and *aflQ* genes. Positive strains were induced to produce aflatoxins on yeast extract sucrose and quantified using liquid chromatography-tandem mass spectrometry (LCMSMS). *Aspergillus flavus* was the most common contaminant and the incidence of occurrence in Nandi and Makueni was not significantly different (82.33% and 73.26%, respectively). Toxigenic strains were more prevalent than non-toxigenic strains. All the toxigenic strains from Makueni were of the *S*-type while those from Nandi belonged to the L-type. Quantitative differences in aflatoxin production *in vitro* between isolates and between strains were detected with S strains producing relatively larger amounts of total aflatoxins, B toxins and lower values for G toxins. This was in accord with the frequent aflatoxicosis outbreaks in Makueni. However some L strains produced considerable amounts of B toxins. Given the widespread distribution of toxigenic strains in both regions, the risk of aflatoxin poisoning is high when favourable conditions for toxin production occur.

INTRODUCTION

Maize is a staple food crop in Kenya for both urban and rural areas with an estimated 1.6 million ha under cultivation. It is grown by both small- and large-scale farmers. Small-scale farmers contribute 75% of the total maize produced in the country. The quantity of maize consumed in Kenya per person per year is approximately 98 kg translating to about 30–34 million bags (1 bag = 90 kg) per year, and in some families maize may be consumed twice daily. Each family in Kenya has a garden if not a farm where they grow maize that is mostly for their own consumption and sometimes for sale signifying the importance of this crop in the country.

Several cases of aflatoxicosis have been reported in Kenya yearly since 1981 following consumption of maize contaminated with *Aspergillus flavus* and aflatoxins. In 1981 the outbreak was as a result of drought followed by heavy rains during harvest of home grown maize. The worst outbreak was in 2004 where 317 cases and 125 deaths were reported, and ever since cases have been reported yearly. In 2010 Kenya had about 2.3 million bags (estimated at $69 million) of maize contaminated with aflatoxins making it unfit for both human and livestock consumption and also for trade, which was a loss to the small-scale farmers who depend on the crop for food and income.

Aspergillus flavus and *A. parasiticus* are ubiquitous and cosmopolitan fungi producing aflatoxins on a wide variety of substrates such as maize, peanut and cotton. *Aspergillus flavus* is a very important toxigenic fungus that produces aflatoxins that are toxic and of great concern because of their health effects on humans and animals. *Aspergillus* can attack crops at different times, in the field, during harvest, transport and storage. High moisture and temperatures are favorable for the growth of this fungus and toxin production. In this study two agro-ecological zones were selected to compare the distribution in maize of *Aspergillus* spp. and their toxigenicity. Nandi County was chosen because it is the main maize growing zone in the country, but no aflatoxicosis has been reported here. The other zone is Makueni County where most aflatoxicosis cases have been reported.

MATERIALS AND METHODS

Three sub-locations were sampled from each of County. Shelled maize were collected from smallholder farm storage facilities of selected households. The grains were analyzed for the presence of fungi by subjecting them to surface sterilization, then plating on differential culture media and identifying them (Klich, 2002). Morphologically identified *A. flavus* and *A. parasiticus* were tested for the presence of aflatoxigenic genes *aflD* and *aflQ* by extracting DNA then running PCR. Aflatoxin production ability was tested and quantified for *A. flavus* and *A. parasiticus* randomly selected from each of the collected samples in order to determine both the fungi that produce great quantities of aflatoxins and the frequency of occurrence of non-aflatoxin producers in Nandi and Makueni counties. To test the ability of toxigenic *Aspergillus* species to produce aflatoxins, isolates were cultured on coconut cream agar and incubated unilluminated at 28°C for 3 days then observed under UV light for reverse fluorescence. To quantify and describe the aflatoxin produced, the toxigenic isolates were cultured on yeast agar to induce production of aflatoxin. After 5 days of unilluminated incubation at 28°C, the mycelia were harvested from the colonies, weighed (5 g) and placed in polystyrene tubes (50 ml) to which 25 ml 70% methanol was added in the ratios of 1:5. The tubes were put on the shaker for 60 minutes, after which the extract was filtered through 0.45 μm and analyzed by Liquid Chromatography Mass Spectrometry coupled with Mass Spectrometry (LCMSMS) on a Waters API Xevo™ coupled to a Waters UPLC.

RESULTS

Population composition of the genus *Aspergillus*

Aspergillus species were the most predominant fungi identified on maize followed by species belonging to the genera *Fusarium* and *Trichoderma* while *Penicillium* and *Rhizopus* were the least dominant. The following *Aspergillus* species were isolated from maize: *A. flavus. A. parasiticus, A. tamari, A. aculeatus, A. carbonarius, A. japonicus, A. fumigatus, A. niger* and *A. ochraceus*. Members of Section *Flavi*, identified with a bright orange reverse colour on the Aspergillus flavus parasiticus agar (AFPA) medium (Figure 1a), were the most dominant species occurring in maize constituting 60% of the total isolates. This was followed by Section *Nigri* (27%), Section *Fumigati* (9.7%), Section *Circumdati* (3%) and Section *Clavati* (0.3%), Negative isolates had either cream or black reverse colours as shown in Figure 1b.

Within the locations the highest incidence of Section *Flavi* was recorded in Nandi and the lowest in Makueni (Table 1). Three species, *A. tamarii, A. flavus* and *A. parasiticus* were all recovered from maize kernels. *Aspergillus flavus* was the most prevalent in both Nandi and Makueni (58.3%; n = 78; 82.8%; n = 87, respectively), followed by *A. tamarii* (25.8%; $n = 33$; 10.8%; $n = 10$) and *A. parasiticus* (15.9%; $n = 21$; 6.6%; $n = 6$).

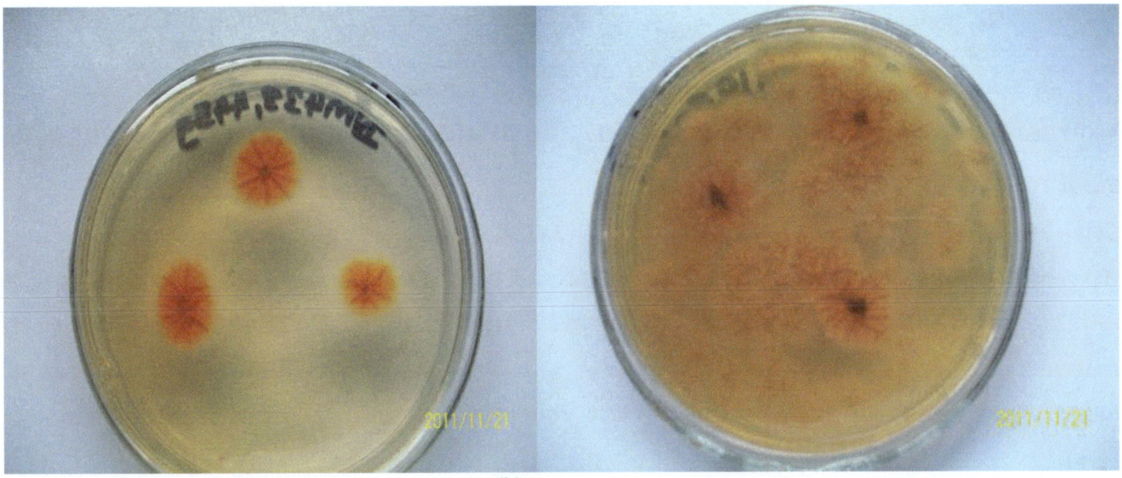

(b)

Figure 1. *Aspergillus* isolates grown on *Aspergillus flavus parasiticus* agar (AFPA) medium to identify members belonging to Section *Flavi*; (a) Positive isolate, (b) Negative isolate

Morphological characteristics of *A. flavus* and *A. parasiticus* on AFPA and V8-juice was used to differentiate the two species. *Aspergillus flavus* had yellow-green colonies, was predominantly biseriate or biseriate and uniseriate with smooth to finely rough conidia; while *A. parasiticus* had dark green colonies with a diameter of between 24–36 mm, was predominantly uniseriate or uniseriate with 20% biseriate and had rough conidia. *Aspergillus tamarii* had dark brown colonies with a diameter of 2–10 mm, was uniseriate with spiky globose conidia. The reverse of *A. tamarii* was bright orange in early stages and turned dark brown after four days.

All the isolates were grown on V8-juice agar for sclerotia formation. Sclerotia formed on the surface of the agar after 3–7 days of inoculation. Sclerotia formation was more frequent from Makueni *A. flavus* isolates (71%; n = 61) compared to Nandi (37%; n = 29). Further, Makueni isolates were all of the S-type producing sclerotia with a diameter range of 200–320 µm. Nandi isolates on the other hand were all L-type producing sclerotia of a diameter range of 1000–3100 µm. Sclerotia formed by *A. parasiticus* isolates from Nandi (935–1590 µm) were larger than those of Makueni isolates (240–280 µm). *Aspergillus tamarrii* did not form sclerotia on V8-juice agar (Table 1).

Table 1. Distribution and characteristics of *Aspergillus* section *Flavi* in nandi and Makueni counties

County	Location	Number of isolates	*A. flavus* (%)	Strain	*A. parasiticus*	*A. tamarii*
Nandi	Kilibwoni	33	64	L	21	15
	Kaptumo	65	55	L	9	36
	Laboret	34	62	L	26	12
	No. producing sclerotia (%)		37 (n=29)		19 (n=4)	0
Makueni	Nguumo	41	95	S	5	0
	Ukia	42	67	S	7	26
	Wote	20	94	S	6	0
	No. producing sclerotia (%)		71 (n=61)		67 (n=4)	0

Molecular analysis of *Aspergillus* strains and aflatoxin production on cococnut agar medium (CAM)

Of the 78 *A. flavus* isolates tested in Nandi 71% (n = 55) were positive for the aflatoxigenic genes while in Makueni 87 isolates were tested and 62% (n = 54) positive. This difference was not significant ($\chi^2 0.05(1)$ = 1.2422; p = 0.265). Positive (toxigenic) strains were significantly predominant over negative (non-toxigenic) strains in all the locations. Of the seven *A. parasiticus* isolates from Makueni, four expressed the toxigenic genes while in Nandi out of the 22 isolates, four were positive.

Figure 2 is a representative of the electrophoretic band patterns obtained for both *afl*D (400 bps) and *afl*Q (719 bps). All isolates that were positive for any one or both gene amplicons fluoresced on coconut agar medium (CAM) after 3 days of incubation at 28 °C, whereas those that were negative for both *afl*D and *afl*Q amplicons did not fluoresce. Figure 3 shows colonies of non-aflatoxigenic and aflatoxigenic isolates of *A. flavus* observed under UV light. Figure 4 shows the distribution of toxigenic and non-toxigenic isolates within the locations. Ukia, Laboret and Kaptumo had over 70% occurrence of toxigenic isolates compared with non-toxigenic isolates.

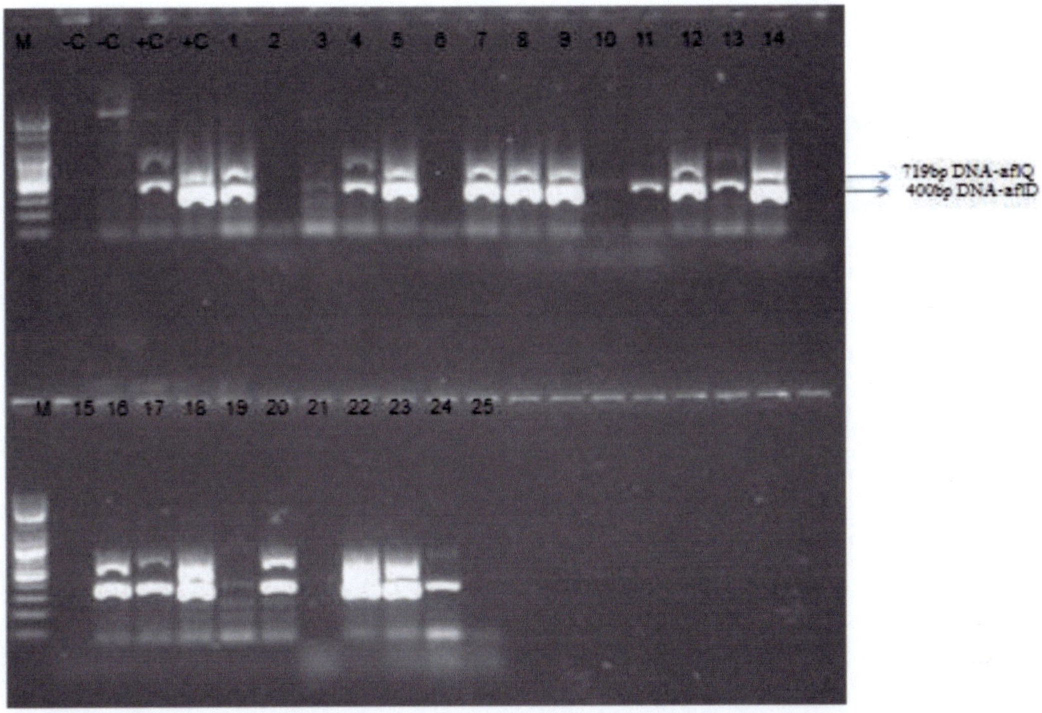

Figure 2. Agarose gel electrophoresis pattern of PCR products showing presence of *aflQ* and *aflD* genes

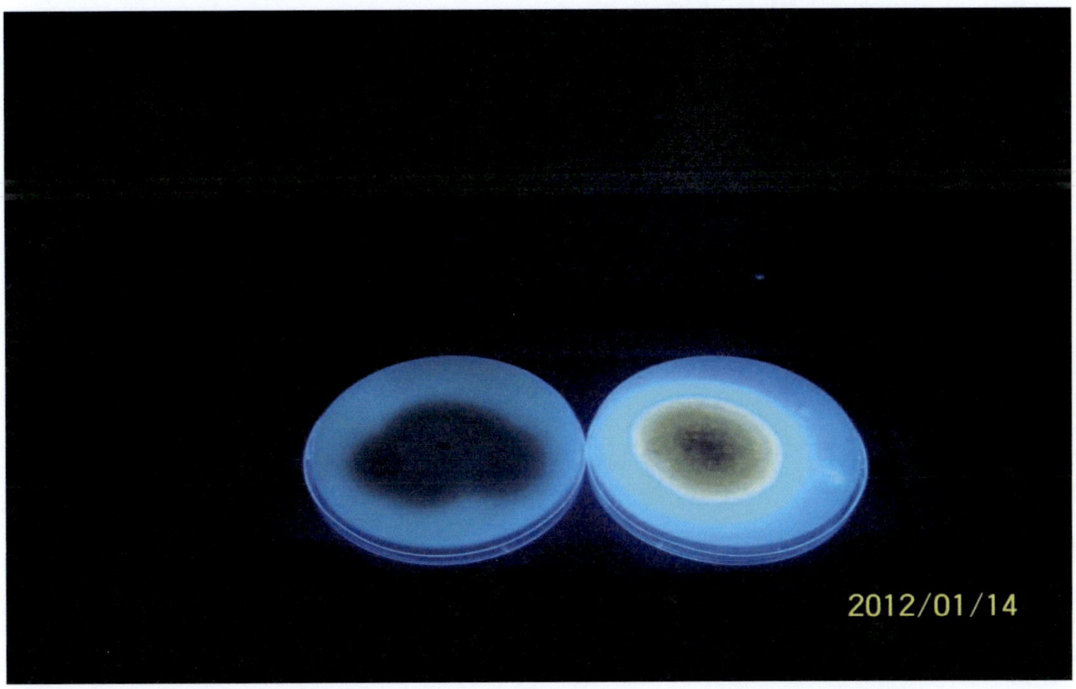

Figure 3. Colonies of non-toxigenic and toxigenic strains of *Aspergillus flavus* observed under UV light after 3 days of incubation on coconut agar medium (CAM).

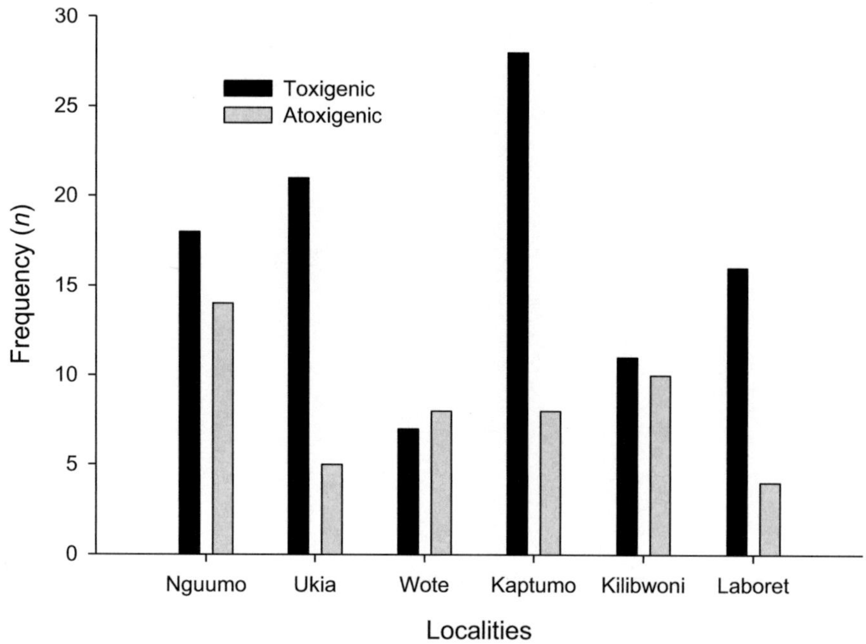

Figure 4. Frequency of toxigenic strains of *Aspergillus flavus* in Makueni county (i.e. Nguumo, Ukia, and Wote locations) and Nandi county (i.e. Kaptumo, Kilibwoni,and laboret locations) in Kenya.

Production of aflatoxins by isolates *in vitro*

All the 78 isolates with toxigenic genes produced AFB1. Only six did not produce AFB2. AFB1 was produced in the largest quantities (highest of 83,491 ppb from Makueni and 64,030 ppb from Nandi) followed by AFB2 (highest 52,600 ppb from Makueni and 59,231 ppb from Nandi), AFG1 (highest 28,728 from Makueni and 48,914 ppb from Nandi) and lastly AFG2 (highest 8,411 ppb from Makueni and 9,629 ppb from Nandi). Makueni isolates produced more AFB1 and AFB2 than Nandi isolates while the reverse was true for the G aflatoxins.

Table 2. Production of aflatoxins *in vitro* by *Aspergillus flavus*

Makueni isolates

Household ID	Aflatoxin B_1 (ppb)	Aflatoxin B_2 (ppb)	Aflatoxin G_1 (ppb)	Aflatoxin G_2 (ppb)
201365	83,491	52,600	16,875	0
201002	63,390	33,777	17	0
100130	61,878	30,349	22	0
100095	59,374	29,455	25	23

Nandi isolates

Household ID	Aflatoxin B_1 (ppb)	Aflatoxin B_2 (ppb)	Aflatoxin G_1 (ppb)	Aflatoxin G_2 (ppb)
BM130	64,030	16,583	51	0
BM1	57,142	59,231	292	0
BM112	13,775	2054	48,914	9,624
BM14	3,771	2,488	9,826	7,798

DISCUSSION

The results obtained in this study have provided, for the first time, an important comparison of fungal contamination between maize sampled in Makueni county, the main region that has experienced repeated lethal human aflatoxicosis outbreaks and Nandi county, the main maize growing region in Kenya. The study shows that *Aspergillus*, specifically Section *Flavi* are the main contaminants of maize in household storage in the two regions and *A. flavus* was the most common species. The incidence of occurrence of *A. flavus* in Nandi and Makueni was the same regardless of the differences in mean temperatures (20°C and 24°C, respectively) and rainfall (900–1800 and 950–1500 mm). High temperatures and drier conditions are known to favour infection by *A. flavus* but this was not the case in this study. Similar results were observed in Nigeria (Atehnkeng et al., 2008).

Toxigenic strains of *A. flavus* were more prevalent than non-toxigenic strains across five out of the six locations. This indicates the risk of aflatoxin poisoning in the event that favorable conditions occur in both regions. The widespread occurrence of the fungus indicates the extent of pre-harvest infection; thus field management strategies stand out as an indispensible intervention strategy towards the fight against aflatoxin contamination of maize in Kenya. In this regard, Abbas et al. (2004) observed a higher incidence and greater numbers of *A. flavus* infection and toxin production when there was no crop rotation. Maize is planted in the same fields every season in the two regions. The importance of field management is further stressed by Zablotowicz et al. (2007) who report that the history of maize cultivation in terms of soil fertility factors correlates with the occurrence of *A. flavus* and toxin production.

All the *A. flavus* toxigenic strains from Makueni maize were of the S-type while those from Nandi belonged to the L type. Quantitative and qualitative differences in aflatoxin production *in vitro* between isolates and between these strains were detected. The S strains were confirmed to produce relative larger amounts of total aflatoxins, AFB1 and AFB2 and lower values for AFG1 and AFG2 *in vitro* compared with the L strains. AFB1 is known to be more toxic than the other aflatoxins, and this explains why the S strain has been associated with acute aflatoxin poisoning in Makueni (Probst et al., 2011). This is accentuated by high temperatures in Makueni (20–28°C) compared to Nandi (18–24°C), which promotes toxin production. However, some L strain isolates from Nandi produced large amounts of AFB1 and AFB2 contrary to the findings of Cotty (1989) and Egel et al. (1994). It is important that *in vivo* tests for toxin production with the L strains from Nandi are done to confirm the capability of these isolates to produce large amounts of toxin. These L strains pose a threat of endemic chronic exposure to humans if the maize is exposed to conditions suitable for toxin production given that they are distributed widely in the region. Further, Nandi is the major maize production zone in the country implying that the maize is distributed to most parts of the country. Control of moisture and temperature during transportation and storage is important since the maize is already contaminated with toxigenic *A. flavus*. The most toxic AFB1 were produced in larger quantities compared to the AFB2. AFG1 were produced in low quantities and only eight out of 78 samples tested produced AFG2.

This is the first report of *A. flavus* S and L strains isolated from Kenya producing both B and G aflatoxins on Yeast Extract Sucrose agar. Probst et al. (2007) isolated in Kenya both S and L strains, which produced only B aflatoxins on maize. Ehrlich et al. (2004) reported that all members of *A. flavus* lack the ability to synthesize G aflatoxins due to a 0.8- to 1.5-kb deletion in the 28-gene aflatoxin biosynthesis cluster agreeing with Cotty and Cardwell (1989) and Egel et al. (1994). However, S strains from Benin produced both AFB1 and AFG1 aflatoxins (Cardwell and Cotty, 2002). Davis et al. (1996) also observed production of aflatoxin AFB1 and AFG1 by *A. flavus* in a synthetic medium. Cardwell and Cotty (2002) reported that SBG strains from the United States and West Africa produced both B and the G aflatoxins. It is not clear if the difference in media is the reason for the variation in the production of B and G toxins or if the isolates that produced both toxins in this study are similar to the SBG strains referred to above. *Aspergillus parasiticus* was rare in both regions. The number of isolates was too small to allow a conclusion on their toxin production ability. The main source of aflatoxin contamination in maize in eastern and western Kenya is *A. flavus*.

Incidences of sclerotia formation were greater among the S-type than the L-type on V8-juice agar. Sclerotia are important survival structures in the life cycle of many fungi. When conditions are favorable, they germinate into hyphae, which then form conidia. The conidia are blown away by the wind and reinfect maize kernels through the silk. Studies on the conditions responsible for sclerotium initiation might be important to develop methods for suppressing the formation of sclerotia, resulting in reduced survival of the fungus and better disease management. Sclerotia have been associated with aflatoxin production (Horn and Dorne, 2002). However the failure of some isolates of *A. flavus* to produce sclerotia on culture media can be due to one of the following: an attenuation of sclerotial production in culture, an unfavorable medium, unfavorable temperature, the differential sensitivity of isolates to light, or other environmental constraints in culture and that strain L isolates require more precise conditions to produce sclerotia than strain S isolates. However, disagreements between studies correlating the sclerotial production of isolates with aflatoxin production exist.

Apart from the risk of aflatoxin contamination of maize, bio-deterioration is another problem associated with high fungal contamination of kernels. The internal mycoflora of maize in Nandi and Makueni is similar and is dominated by the species of *Aspergillus*, *Fusarium* and *Penicillium,* which predisposes the kernels to bio-deterioration. The resulting physiological and biochemical changes in the maize kernels eventually render grains unsuitable for human consumption. In addition, species of the genera *Fusarium*, *Penicillium* and *Alternaria* were isolated and some species are capable of producing a wide spectrum of compounds shown to be toxic to man and animals (Orsi et al., 2000) and so increase the risk of multi-mycotoxin contamination and exposure.

CONCLUSION

This research has shown that *A. flavus* is the main fungal species infecting maize grains in Kenya and that toxigenic strains are widespread. The situation is accentuated in Makueni county by the existence of the highly toxigenic S strain and high prevailing temperatures. However, the presence of the L strains should not be overlooked, as this indicates the possibility of chronic exposure from the lower levels of toxins produced by these strains. The influence of aflatoxins on human populations in Kenya over the past decade demonstrates a clear need for tools to manage contamination of locally produced maize. Given the widespread nature of toxigenic strains, any control strategy will have to include field interventions. Fungal infection of maize varies with households and this reflects different farm management practices, some of which adversely contribute to contamination of the maize value chain.

ACKNOWLEDGEMENTS

This work is part of the Safe Food Safe Dairy Project funded by the Finnish Ministry of Foreign Affairs. We also extend our gratitude to African Women in Agricultural Research and Development (AWARD) for their financial support for toxin analysis at the *Fusarium* Laboratory, Department of Plant Pathology, Stellenbosch University in South Africa and the latter for their time and scientific guidance.

REFERENCES

Abbas HK, Zablotowicz RM, Locke MA. (2004). Spatial variability of *Aspergillus flavus* soil populations under different crops and corn grain infestation and mycotoxins. Canadian Journal of Botany, 82:1768–1775.

Atehnkeng J, Ojiambo PS, Donner M, Ikotun T, Sikora AR, Cotty PJ, Bandyopadhyay R. (2008). Distribution and toxigenicity of Aspergillus species isolated from maize kernels from three agro-ecological zones in Nigeria. International Journal of Food Microbiology, 122:74-84.

Cardwell KF, Cotty PJ. (2002). Distribution of *Aspergillus* section *Flavi* among field soils from the four agroecological zones of the Republic of Benin, West Africa. Plant Disease, 86 (4):434–439

Cotty PJ. (1989). Virulence and cultural characteristics of two *Aspergillus flavus* strains pathogenic on cotton. Phytopathology, 79(7):808–814.

Davis ND, Diener UL, Eldridge DW. (1966). Production of aflatoxins B1 and G1 by *Aspergillus flavus* in a semisynthetic medium. Applied and Environmental Microbiology; 14 (3):378–380.

Egel DS, Cotty PJ, Elias KS. (1994). Relationships among isolates of *Aspergillus* section *Flavi* that vary in aflatoxin production. Phytopathology, 84:96–912.

Ehrlich KC, Chang PK, Yu J, Cotty PJ. (2004). Aflatoxin biosynthesis cluster gene *cypA* is required for G aflatoxin formation. Applied and Environmental Microbiology, 70:6518–6524.

Horn BW, Dorne, JW. (2002). Effect of competition and adverse culture conditions on aflatoxin production by *Aspergillus flavus* through successive generations. Mycologia, 94 (5):741–751.

Klich MA. (2002). Biogeography of Aspergillus species in soil and litter. Mycologia, 94:21-27.

Orsi RB, Corrêa B, Possi CR, Schammass EA, Nogueira JR, Dias SMC, Malozzi MAB. (2000). Mycoflora and occurrence of fumonisins in freshly harvested and stored hybrid maize. Journal of Stored Products Research, 36:75–87.

Probst C, Bandyopadhyay R, Price LE, Cotty PJ. (2011). Identification of non-toxigenic Aspergillus flavus isolates to reduce aflatoxin contamination of maize in Kenya. Plant Disease, 95:212-218.

Probst C, Njapau H, Cotty PJ. (2007). Outbreak of an acute aflatoxicosis in Kenya in 2004: identification of the causal agent. Applied and Environmental Microbiology, 73:2762–2764.

Zablotowicz RM, Abbas HK, Locke MA. (2007). Population ecology of *Aspergillus flavus* associated with Mississippi Delta Soils. Food Additives and Contaminants, 24:1102–1108.

14. Toxicity and repellant potency of bush mint (*Hyptis spicigera*) extracts on maize weevil (*Sitophilus zeamais*)

Ingrid Wekesa[1,4*], Leonzio Onek[1], Arop Deng[2], Ahmed Hasanali[3] and Jacktone Othira[1]
[1]Department of Biochemistry and Molecular Biology, Egerton University, Egerton.
[2]Department of Biological Sciences, Egerton University, Egerton.
[3]Department of Chemistry, Kenyatta University, Nairobi.
[4]Chemical Engineering Division, Kenya Industrial Research and Development Institute, Nairobi.
[*]*Corresponding author: wngrid@yahoo.com*

SUMMARY

Bush mint (*Hyptis spicigera*) also known as 'latek' by communities in Northern Uganda, is used as a cereal and legume protectant in control of storage pests as part of their ethnoknowledge. This study focused on determining the toxicity and repellent potencies of *H. spicigera* against maize weevil, *Sitophilus zeamais*. Essential oils, methanol extracts and hexane extracts were tested against maize weevil in fumigation, contact and repellency laboratory bioassays. Results revealed that *H. spicigera* extracts exerted typical toxicity effects on the maize weevil. Notably, essential oils possessed potent dose–dependent fumigant toxicity against *S. zeamais*. Total mortality was noted within 48 hours of exposure in >20 µl essential oil treatments. LC_{50} value was 48.11 µl. Contact assays showed significant mortality of the insects, over 50% after 72 hours and 48 hours of exposure, for hexane and methanol extracts respectively, applied at 50 mg, with decreased progeny emergence. Essential oils and solvent extracts caused over 50% repellency, with 100 µl evoking 84% repellency (RD_{50} 44.32 µl). Analysis of essential oils by GC-MS revealed abundance of mono- and sesquiterpenoids, with constituents identified as a-pinene, β-pinene, a-phellandrene, isocaryophyllene, limonene-co-eluting with phellandrene and L-phellandrene co-eluting with a-thujene. The plant may be exploited for the development of botanical insecticides to be used in weevil management.

INTRODUCTION

Maize is an important component in the diet of Kenyans. To ensure its continued cultivation, pre- and post harvest losses due to agricultural pests must be minimized. According to Tegemeo Institute of Agricultural Policy and Development projections for 2009/2010, Kenya expected maize production of about 27 million bags (1 bag = 90 kg) of maize, against consumption of 37 million bags (Olwande, 2009). The deficit (14 million bags) was met through imports and cross border trade. The deficit could be attributed to unpredictable rain patterns, which are patchy, poorly distributed and insufficient in most parts of the country. This results in most of the farmers being unable to plant and the little that the farmers harvest is subject to attack by mainly insects.

The maize weevil, *Sitophilus zeamais* Motschulsky, was identified as the major insect pest of maize in storage in Kenya (Ogendo, 2000). Probably the most common method of controlling insect pests of maize in storage is the use of synthetic insecticides, owing to their efficiency and rapid action. According to the Pest Control Board (2004), most of the registered stored grain protectants for use in Kenya are organophosphates, pyrethroids and pyrethrin based. Though pesticides are considered as the silver bullet in insect control, problems keep arising that are associated with their continued use. Some of the potential risks highlighted are environmental degradation, insect resistance and mammalian toxicity.

To control insect pests, various compounds which disturb neuronal regulation including those acting on cholinergic system have been developed. Cholinesterase–inhibitor organophosphorus compounds have been in use since 1936 (Gyori et al., 2007). The use of methyl bromide, an effective fumigant against stored products pests and soil pests, was to be reduced by 20% in 2005 and 100% restricted in 2015 in developing countries, due to its potential ozone depleting properties; exempt were specific uses such as pre-shipment, quarantine and emergency use (Ristaino, 1998). Insecticide resistance is an evolutionary response to insecticides and as such, important for environmental biomonitoring and for pest management (Oliveira et al., 2007). Different levels of insecticide resistance, reduced susceptibility and behavioral avoidance have been reported in populations of *S. zeamais* (Ribiero et al., 2003; Braga et al., 2011). Potential genotoxicity, immunotoxicity and effects on the reproductive system on mammals (Ibrahim and El-Gamal, 2003; Sanchez-Pena et al., 2004; Lecoeur et al., 2006) associated with the continued use of pesticides in insect control has been demontsrated. The carcinogenic potential of some pesticides to humans has also been noted (Fishel, 2005).

Heavy reliance on chemical pesticides in preventing pre- and post-harvest losses, caused by insect attack is not viable as they provide ephemeral benefits often associated with adverse side effects. Prohibitive costs associated with pesticide use also make it unattractive to resource poor farmers. This has become a major concern and thus given impetus to search for alternative methods of pest control (Obeng-Ofori and Amitaye, 2005). The advent of synthetic insecticides in the protection of stored products and their associated drawbacks has refocused attention to ancient practices used in post-harvest protection. Botanicals are one of the most promising alternative pest control methods. Especially in tropical regions, the application of botanical material to protect a crop against insects is often traditional and centuries old (Boeke et al., 2004). Botanicals currently used as insecticides constitute 1% of the world insecticide market (Rozman et al., 2007). This is despite the knowledge that plants constitute a rich source of bioactive chemicals and may provide alternatives to currently used insect control agents (Kim et al., 2003). They are attractive candidates for use in integrated pest management since they are active against a limited number of species, including specific target pests and are often biodegradable to non-toxic products (Soon-II et al., 2003). Regardless of the cosmopolitan use of plants as grain protectants, knowledge of methods and efficacy of these plants is scanty. Most of the accessible information is derived from anecdotal sources, which provide minimal information (FAO, 1998).

Various plant extracts have been evaluated in laboratory studies for their ability to protect stored maize against insect pests mainly *Sitophilus* species, *Prostephanus truncatus, Rhyzopertha dominica* and *Tribolium confusum*. Essential oils from *Eucalyptus saligna* and *Cupress sempervirens* have been shown to have repellent and toxic effects on *Sitophilus zeamais* and *Tribolium confusum. Eucalyptus saligna* essential oil was found to be more toxic than *Cupress* oil to both insect species on filter paper disks. Both oils were found to considerably reduce F_1 progeny production and grain weight loss (Tapondjou et al., 2005). In addition, both crude oil extracts produced stronger repellent activity than cymol (one of their major oil constituents). Investigations by Chu et al. (2010) demonstrated that the essential oil of *A. vestita* possessed strong fumigant toxicity against *S. zeamais* adults with a LC50 value of 13.42 mg/L air. The essential oil of *A. vestita* also showed contact toxicity against *S. zeamais* adults with a LD50 value of 50.62 mg/adult. The main components of the essential oil were identified as grandisol (40.29%), 1,8-cineol (14.88%) and camphor (11.37%) (Chu et al., 2010).

Studies by Zoubiri and Baaliouame (2011) demonstrated the fumigant activity of *L. citriodora, F. vulgare* and *R. officinalis* essential oils against *S. granarius*. In their study, fumigation for 120 h for oil concentrations at 5 µL showed 50% mortality of *S. granarius* for the three plants oils with complete control being achieved by treating air with 50 µL of oil. Essential oils of *X. aethiopica,* when incorporated in clay caused mortality in the maize weevil according to Nguemtchouin et al. (2010). The insecticidal formulation produced mortality rates ranging from 22% at 2.5% (w/w) to 100% at 10% (w/w). The lethal concentration was established as 4.35% (w/w). The efficacy was found to be dependent on storage conditions and duration of the formulation (Nguemtchouin et al., 2010).

Powders of *Ocimum americanum* have been shown to dose dependently increase the mortality and F_1 emergence of *P. truncatus*, *S. zeamais* and *Callosobruchus maculatus*, when included in 100 g maize at 0.5, 1.0 and 5% (Belmain et al., 2001). *Chemopodium ambrosiodes* powders and essential oils from leaves were evaluated for their efficacy as post-harvest grain protectants. Powders applied at 6.4% w/w induced total mortality in *S. zeamais* and *S. granarius* 48 h after treatment. In addition, application of 0.2 $\mu l/cm^2$ essential oil killed 80-100% of the beetles (*Callosobruchus chinenesis, Acanthoscelides obtectus, S. granarius and P. truncatus*) except *C. maculatus* and *S. zeamais*, where this concentration induced only 20% and 5% mortality respectively, in laboratory studies (Tapondjou et al., 2002).

Role of *Hyptis spicigera* in pest management

Aromatic plant species, particularly those in the family Labiatae, are among the most widely used plants in insect pest control (Ke'ita et al., 2000) and are backed with experiential evidence (Belmain et al., 2001; Ogendo et al., 2004). They are among the most efficient insecticides of botanical origin and essential oils often constitute the bioactive fraction of the plant extracts (Cosimi et al., 2009). Of particular interest is Bush mint (*Hyptis spicigera*), which is claimed to have insecticidal effects against pests by communities in Southern Sudan and Northern Uganda. They incorporate it in legumes and cereals during post-harvest storage and also burn whole plants of the same to repel mosquitoes (Onek et al., 2003).

Few laboratory studies have been undertaken to demonstrate the insecticidal potential of *H. spicigera*. Powders of *H. spicigera* were investigated for their potential use as stored grain protectants against *Sitophilus oryzae* and *Prostephanus truncatus* when used as direct admixtures to the grain they resulted in mortalities of the two insect species (Niber, 1994). When tested on the European corn borer larvae, *H. spicigera* was found to exert insecticidal activity. A fraction containing a rich mixture of new labdane diterpenes was responsible for this activity (Rogelio et al., 2001). Other plants in the same genus (*Hyptis*) shown to have insecticidal activity include *H. verticillata* which possessed insecticidal activity against the sweet potato weevil (Buchanan et al., 2000) and *H. suaveolens*, which was found to exert ovicidal activity in *C. maculatus* (Ke'ita et al., 2000). In repellency studies of African plants used in the protection of stored cowpea, *H. suaveolens* was found to be attractive to *C. maculatus* (Boeke et al., 2004).

Hyptis spicigera taxonomy and uses

The genus *Hyptis* belongs to the mint family Labiatae. It is predominantly herbaceous, having only a few species which are half shrub (woody only at the base). *Hyptis Spicigera* is an erect aromatic herb which grows up to 1 m in height. It produces terminal inflorescence, which are dense and cylindrical or ovoid cylindrical, spike to 9 cm long, of very small white/mauve flowers. Its habitat stretches from Senegal to west Cameroon. It is possibly native to Brazil, although it is now widely naturalized in tropical Africa and Asia (FAO, 1999). The seeds are eaten whole, ground, cooked in stews or incorporated in peanut butter, among the Acholi of Sudan. They are normally intercropped with maize and simsim in the Northern parts of Uganda and Southern Sudan. Traditional medicinal uses these plants include treatment of gastro-intestinal disturbances, wounds, skin infections and insect bites in traditional Mexican medicine (Rogelio et al., 2001).

Biochemistry of *Hyptis spicigera*

Crude extracts of *H. spicigera* were shown to be cytotoxic, (ED 18 g/ml), *in vitro* on human nasopharyngeal carcinoma assay system. This led to the isolation of spicigerolide (Figure 1) through bioactivity directed fractionation of the crude extracts (Rogelio et al., 2001). Chemical studies on the aerial parts of this plant led to the isolation of spicigera-δ-lactone (Figure 2) from the petroleum extract of inflorescences. Further studies have revealed that it is apparently the same as 5-deacetoxy-5-epiolguine, a cytotoxic constituent of *Hyptis oblongifolia* (Aycard et al., 1993; Rogelio et al., 2001). Other compounds isolated from a plant, *Hyptis pectinata* of the same genus include α-pyrones (Figure 3) and a 2(5H)-furanone (Figure 4) (Boakino et al., 2003). A novel 8.7 KDa protease inhibitor, with potent inhibitory activity towards all trypsin-like protease, extracted from the gut of *P. truncatus* was purified and characterized in biochemical studies on *H suaveolens* seeds (Aguirre et al., 2004).

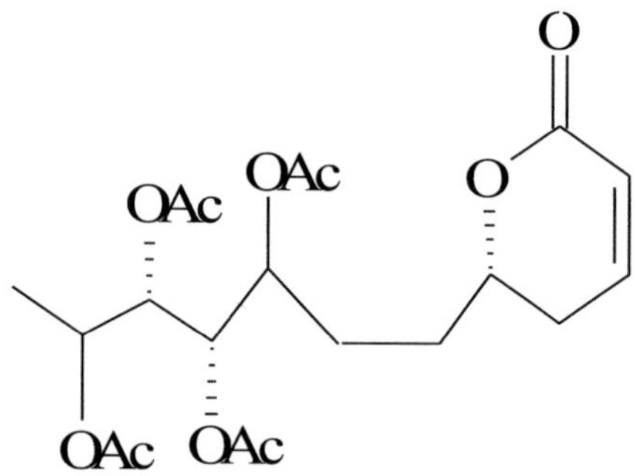

Figure 1. Spicigerolide

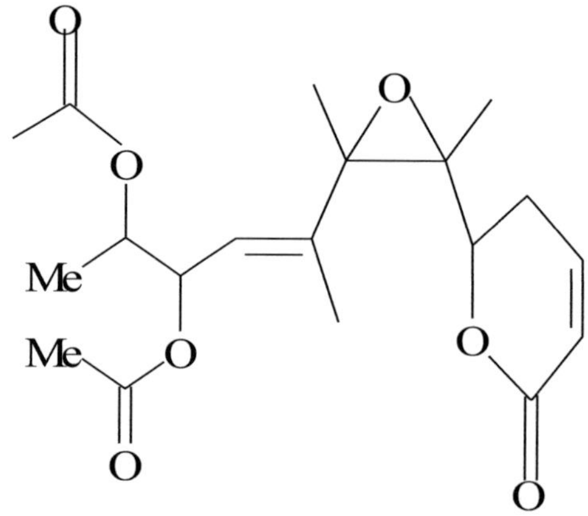

Figure 2. Spicigera-δ-lactone

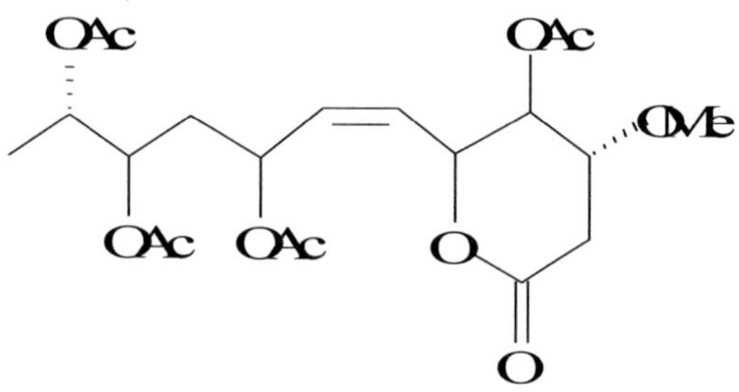

Figure 3. Pyrone

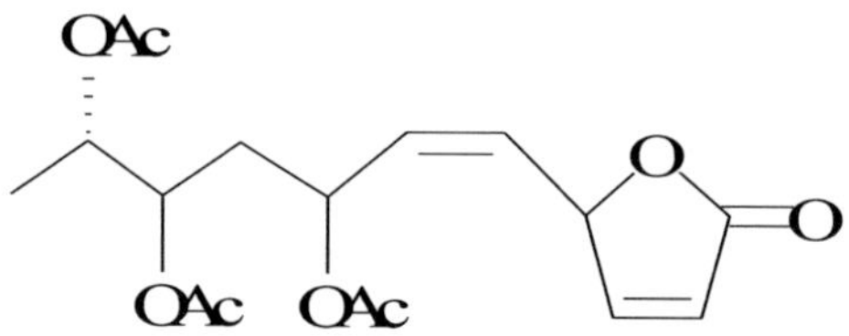

Figure 4. Furanone

As demands increase for sustainable agriculture and concern grows regarding the extensive use of synthetic chemicals, attention is focused on alternative strategies for meeting some of these demands. This seeks to establish whether a scientific rationale exists for the use of *H. spicigera* in anti-weevil ethno-knowledge, as part of its residue management. The ultimate acceptance of *H. spicigera* as a tool to control storage pests will be premised on its other economic benefits to the farmers and its efficacy in the control of storage insect pests provides the best way forward.

Staple food production is inadequate for the ever burgeoning population, making increased crop productivity and preservation of the harvest key in contributing to food security. The loss of food grains due to pests remains a monumental challenge, especially as the feeling of optimism in use of synthetic pesticides wanes, in the wake of manifold adverse effects. Grain production and storage practices are now focused on those which promote 'green chemistry' particularly ethno-knowledge. In many of these practices, the scientific rationale for their use has remained by and large uninvestigated with little or no documentation. This may provide insight into bioactive compounds, which may be incorporated into the current pest management systems, to mitigate effects of insect incidence and attack.

This study employed detailed laboratory investigations geared towards determining the effects of *H. spicigera* extract (leaf powders, solvent extracts and essential oils) on management of *Sitophilus zeamais* in stored grain in an effort to establish the scientific rationale behind its use in traditional storage systems.

MATERIALS AND METHODS

Collection and preparation of plant materials

Hyptis spicigera flowering plants were collected from Ajumani in Northern Uganda. More plant materials were available at ICIPE-Kasarani, where they had been planted in pots in the screen house. *H. spicigera* plants were then dried in the shade (drying room) for 5 days and finely powdered using a blender and stored in air-tight paper bags until when required. Dried leaves and inflorescence of *H. spicigera* were subjected to steam distillation (separately) to obtain essential oils which were collected over hexane. The essential oils were concentrated by blowing off hexane under a stream of nitrogen and stored at 4°C in coloured vials until when required.

Preparation of maize seeds

Good quality infection-free maize seeds, free from insecticides to be used in insecticidal studies, were disinfested by heating in an oven at 40°C for 1 hour and left to cool overnight before use.

Sourcing and rearing of insects

The weevils were reared on maize free from insecticides, isolated from the infested stock after a 7-day egg laying period, and kept at 17-28°C, relative humidity 38-69% to obtain newly emerged weevils of the same population (F_1 generation), which were used for bioassays.

Solvent extracts preparation

Ground leaf samples (50 g) prepared above were extracted twice with 300 ml solvents (methanol, hexane), at room temperature, for 2 days and filtered using Whatman No.2 filter papers. The combined filtrate was concentrated to dryness by rotary evaporation at 40°C. The yield of each solvent extraction was calculated and recorded (Kim et al., 2003).

Fumigant toxicity of *H. spicigera* essential oils

Ten pairs of 3 week old weevils were placed in vials capped with polypropylene stoppers and 5, 10, 20, 30 and 40 µl of essential oil of *H. spicigera* placed in similar vials. Vials containing insects were turned upside down over the vials containing oils, to allow saturation of atmosphere with oil vapours. Controls contained similar set up but without essential oil. The treatments in the bioassays were laid out in a completely randomized design and the procedure was replicated 5 times. Mortality counts were made 24, 48 and 72 hours after treatment. Fumigation procedures were performed on the methanolic and hexanolic extracts in a similar manner. The insects were considered dead if appendages did not move when probed with a camel brush (K'eita et al., 2000).

Contact toxicity of *H. spicigera* solvent extracts

Bioassays involving crude solvent plant extracts of *H. spicigera,* prepared as described above were used in contact toxicity assays according to Kim et al. (2003) with some modifications. Briefly, 20 g maize grains in Petri dishes were coated with 50 mg and 25 mg plant extracts dissolved in 3 ml methanol/hexane. These were then place under a fume hood for 10 min to allow the solvent to evaporate. Ten insect pairs were then introduced into each dish and exposed to treatments. The observation time was 1, 2, 3, 4, 7, 14 and 21 days, after exposure. Controls contained maize grains mixed with 2 ml of either methanol or hexane and maize grains not treated at all. The treatments in the bioassays were laid out in a completely randomized design and replicated 3 times. Percent mortalities were determined. After 21 days the weevils were removed from the treated grains and the grains kept for 60 days after which F_1 progeny was counted and recorded. Percentage mortality (adjusted) was calculated using Abbot's formula (Abbot, 1925).

Repellency tests with leaf powders, solvent extracts and essential oils

Repellent effects of *H. spicigera* extracts against the maize weevil were performed using a choice bioassay system according to Bekele (1995). The system consisted of two 1-litre glass jars connected at their rims by means of a nylon mesh tube. Fifty adult insects were introduced into the system via a circular hole at the middle of the upper side of the nylon mesh. Each jar contained maize seeds; treated with leaf powders (1 g, 4 g and 7 g) or solvent extracts (100 mg, 50 mg) or essential oils (10 µL, 40 µL, 70 µL and 100 µL) while the control jars contained untreated maize seeds. The treatments in the bioassays were laid out in a completely randomized design (CRD) and replicated three to four times. The number of insects present in the control jar (N_C) and the treated jars (N_T) were recorded after 24 hours of exposure. All bioassays were carried out in the laboratory maintained at 26±5°C and 60±5% relative humidity. Percent repellency (PR) values were computed using the method by Hassanali et al. (1990). PR data was analyzed using analysis of variance after having being normalized using $\log_{(10)}(X+1)$ transformation.

Chemical analysis of essential oils

Gas chromatographic (GC) analyses were performed using a HP ultra 1 methyl silicone capillary column (50 m x 0.2 mm, phase thickness 0.33 μm). The temperature program was: initial temperature 60°C for 5 minutes, which was increased to 280°C at 5°C/minute and maintained at 280°C for 20 minutes. Nitrogen was the carrier gas. Sample for injection into the GC was prepared as follows: 1μl of essential oil dissolved in 1ml DCM. 1 μl of this was then injected into GC. For GC-MS analyses, helium (100.0000%) was used as carrier gas and maintained at a flow rate of 1 ml/minute in a split/split less flow. The essential oils constituents of the different oils were identified by comparisons of their mass spectra with those of NIST and WILEY libraries, co-injection with known available authentic standards and use of retention times.

Data analysis

The treatments were laid out in a completely randomized design. The original data (% and counts), were transformed using $\log_{10}(X+1)$ to normalize the variance and subjected to General Linear Model and means separated by DMRT and Tukeys Studentised Range Test at 0.05% level of significance. Probit analysis was used to determine LC_{50} values (Finney, 1978).

RESULTS AND DISCUSSION

Insecticidal potency of *H. spicigera* essential oils

a) Fumigation effect of *H. spicigera* essential oils

The results of fumigation assays of *H. spicigera* essential oils against *S. zeamais* are presented in Table 1. The mortality rate of *S. zeamais* increased with the concentration and duration of exposure to essential oils. Total mortality was obtained within 48 h of exposure for three doses higher than 10 μL of the essential oils tested. The LC_{50} values obtained from Probit analysis are 44.32, 48.11, 46.82 and 46.67 for 24 h, 48 h, 72 h and 5 days respectively (Table 2).

Table 1. Effects of *H. spicigera* essential oils on mortality (mean % mortality ± s.e) of adult *S. zeamais*.

Concentration	Days after exposure			
	1	2	3	5
0 μL	0±0.00^{e}	0±00^{e}	0±0.00^{b}	0±0.00^{b}
10 μL	8.75±0.82^{d}	71.25±6.25^{b}	96.25±2.39^{a}	100±0.00^{a}
20 μL	23.75±1.57^{c}	100±0.00^{a}	100±0.00^{a}	100±0.00^{a}
30 μL	47.50±2.11^{b}	100±0.00^{a}	100±0.00^{a}	100±0.00^{a}
40 μL	83.75±2.80^{a}	100±0.00^{a}	100±0.00^{a}	100±0.00^{a}

Means in columns with the same letter are not significantly different at P > 0.05.

Table 2. Efficacy of essential oil of *H. spicigera* as a fumigant for *S. zeamais*.

Treatment type	$L_{C\text{-}50}$	Class interval
24 hours	44.32 μl	35.07-45.41
48 hours	48.11 μl	42.45-58.14
72 hours	46.82 μl	41.84-54.99
5 days	46.67 μl	41.77-54.64

b) Contact toxicity of *H. spicigera* solvent extracts and progeny emergence

Figure 5 shows the results for bioassays involving direct application of the solvent extracts onto the maize grains. Both extracts at low dose levels (25 mg), did not differ significantly ($P > 0.05$) in mortality values from each other and from the controls at all exposure periods (Figure 5). Hexane and methanol extracts at the highest dose (50 mg) differed significantly ($P \leq 0.05$) from each other in mortality values after the first 5 days of exposure. Nonetheless in both extracts mortality seemed to increase with an increase in the exposure period (Figure 5). Over 50% mortality was achieved after 2 and 3 days of exposure for methanol and hexane extracts respectively. Total mortality was attained after 5 days of exposure to methanol extracts and 14 days after exposure to hexane extracts (Figure 5).

Progeny emergence (F_1) in maize which had been treated with the solvent extracts revealed that significant ($P \leq 0.05$) higher progeny were observed in the controls (no treatment applied, methanol only and hexane only), compared to the solvent extracts, at the different applied doses. However, the emergence in treatments for both hexane and methanol extracts at 25 mg were not significantly ($P > 0.05$) different from each other. On the contrary, methanol extracts at 50 mg had significantly ($P \leq 0.05$) higher progeny than those observed in the hexane treatments at the same applied dose (Figure 6). Generally, all treatments were significantly different from the controls and from each other, with those applied at the lower dose in both extracts not being significantly different from each other. However, there were significant differences between the two higher dose levels of the different extracts.

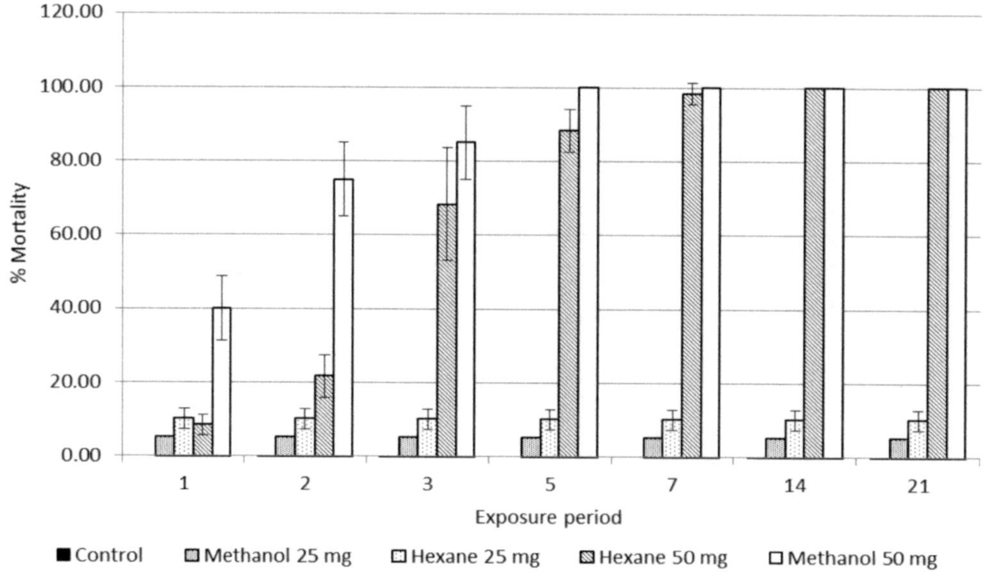

Figure 5. Mean mortality of *S. zeamais* in grains treated with *H. spicigera* solvent extracts.

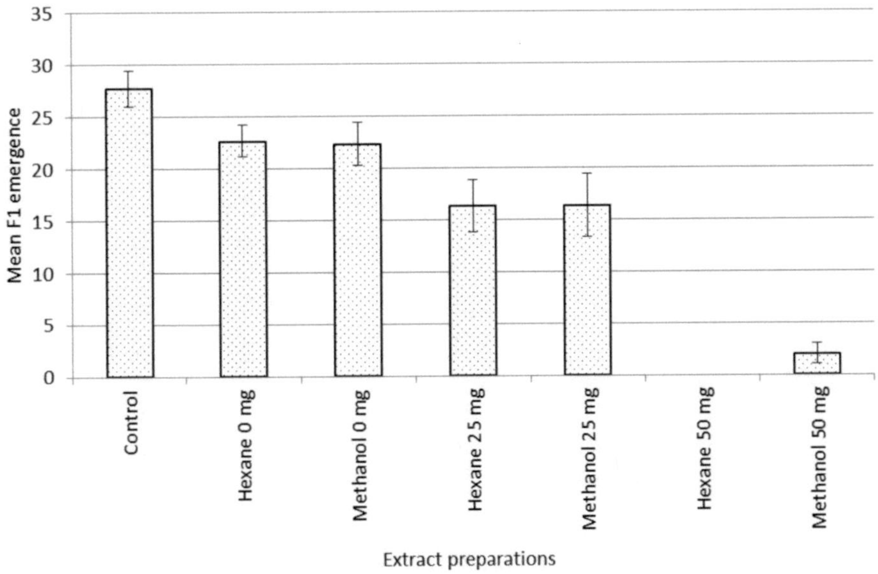

Figure. 6. Mean numbers of *S. zeamais* progeny (F_1) emergence at 60 days after application of *H. spicigera* extracts on maize grains.

c) Repellent activities of *H. spicigera l*eaf powders, solvent extracts and essential oils

Figure 7 shows the mean repellency values for each of the three *H. spicigera* test materials at different doses. Analysis of variance indicated that there were significant ($P \leq 0.05$) differences in repellency studies using powders of the plant at different doses. The repellency of the ground leaf powders appeared to be dose dependent, with 7 g evoking the highest repellent action, though not significantly different from 4 g, within the same exposure duration. On the other hand, repellency at 1 g was significantly ($P \leq 0.05$) different from both 4 g and 7 g treatments.

Solvent extracts were observed to have repellent effect on the maize weevil. The observed repellency of all solvent extracts (50 mg and 100 mg doses) followed a dose-dependent pattern. However, in case of methanol extracts there were no significant differences between the two concentrations tested. On the other hand, hexane extracts at a higher dose (100 mg) showed a significantly ($P \leq 0.05$) higher repellency than methanol extracts at all rates applied and at its lower dose (50 mg).

Essential oils applied at the different doses generally evoked higher repellency to *S. zeamais* than the powders and solvent extracts. The lowest dose (10 μl), repelled more than 50% of the weevils after treatment, while the highest dose (100 μl), induced more than 84% repellency within the same exposure time. The RD_{50} for essential oils was found to be 44.32 μl from Probit analysis.

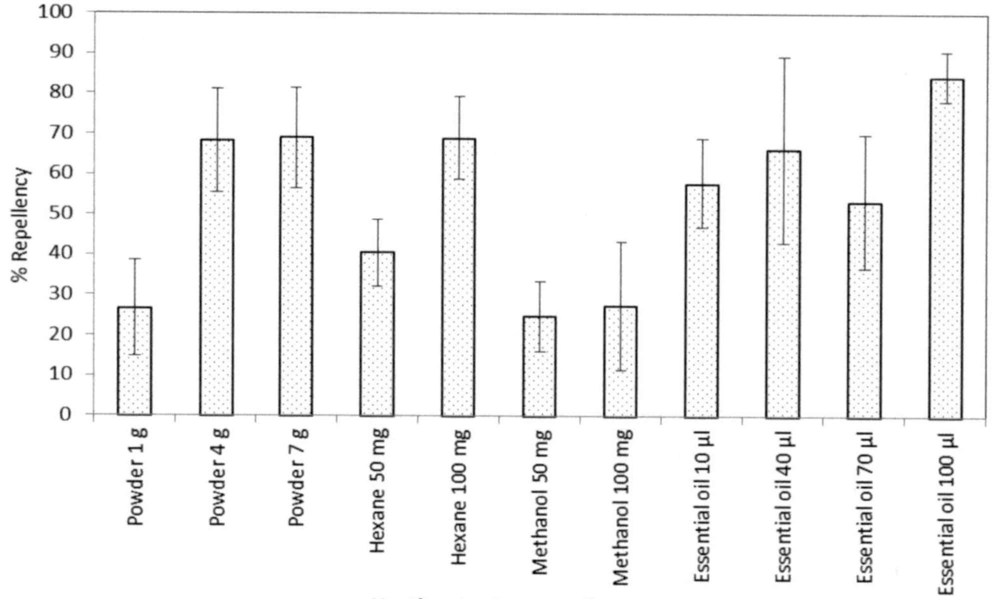

Figure 7. Repellant effects of various preparations of *H. spicigera* extracts on maize grains infested with *S. zeamais*

Chemical constituents of the essential oils

Essential oil yields obtained from dry weighed plant samples as follows: 0.867 g (0.181%) of oils was obtained from 480 g inflorescence and 1.368 g (0.182%) of oil was obtained from 750 g leaves. The oils obtained were pale yellow for both inflorescence and leaves. The yields from hexane and methanol extracts, after rotary evaporation were 1.181 g (2.362%) and 3.2138 g (6.43%) respectively.

Table 4. Major chemical constituents of essential oils from *Hyptis spicigera.*

Compound	% Composition	
	Leaves	Inflorescence
(1, 1') Alpha pinene	24.6	27.4
(2, 2') Beta-phellanderene	3.3	5.3
(3, 3') Beta pinene	13.3	17.6
(4, 4') L-phellandrene/alpha-thujene	6.2	12.2
(5, 5') Limonene/phellandrene	5.5	9.4
(8, 8') Alpha copaene	2.8	1.2
(7, 7') Terpinene-4-ol	1.0	1.0
(9, 9') Bourbonene	2.6	<1%
(10,10') Isocaryophyllene	18.2	9.2
(11,11') Alpha humulene	2.3	1.1
(12,12') Germacrene D	2.5	1.5
{Gamma cadinene	1.3	1.1
(13, 13') {Beta cadinen	<1%	1.1
{Delta cadinene	1.9	<1%
(14, 14')Caryophyllene oxide	1.1	<1%
(15, 15') unknown	2.8	1.9

Further work in the study focused on use of *H. spicigera* residues in management of *S. zeamais*. This involved use of essential oils, solvent extracts and powders *H. spicigera*. Fumigation studies carried out showed that the essential oils of *H. spicigera* had outstanding toxicity at the dose levels tested. Since there was no contact between the treated filter papers discs and the test grains and insects, fumigant action was suggested as the cause of the observed mortality. The essential oils of *H. spicigera* had a 'knock down effect' on the insect pest in fumigation studies. Similar results on fumigant activity were obtained when newly emerged maize weevil adult insects were exposed to *H. spicigera* essential oils, at a dose of 25 μl/vial, for up to 12 hours in fumigation studies recording 70% mortality (Othira et al., 2009). The mode of action of essential oils of *H. spicigera* points towards inhibition of insect acetyl cholinesterase and thus ultimate block in nerve function. This is supported by observations made such as: signs of immobilization with flexed legs and clinging to the grain, outstretched metathorasic wings from the elytra and paralysis of the dead/dying insects in agreement with similar studies by Obeng-Ofori and Amitaye (2005). The observed rapid action of essential oils could be attributed to their property of acting in the vapour phase, hence gaining entry into the insect's internal systems with ease through the spiracles. In topical application procedures, the insect is protected by its exoskeleton against external influences.

Investigation of fumigant toxicity of essential oils from rhizomes of *Alpinia conchigera, Zingiber zerumbet, Curcuma zedoaria* and their major compounds against adults of *Sitophilus zeamais, Tribolium castaneum, Anisopteromalus calandrae and Trichogramma deion larvae* was studied by Suthisut et al., (2011). The study demonstrated the toxicity of *Alpinia conchigera* oils to *S. zeamais, T. castaneum and T. deion,* with the other two plant oils recording low toxicity. Adults of *S. zeamais* and *T. castaneum* were more susceptible to *A. conchigera* oils than their eggs, larvae or pupae. The study also revealed that synthetic essential oils were more toxic than the extracted essential oils to both insects. *Tribolium castaneum* was more susceptible than *S. zeamais* to the eight pure compounds. Terpinen-4-ol was highly toxic to both insects (Suthisut et al., 2011).

Maize treated with hexane and methanol extracts (25 mg) had low insect mortality numbers being obtained, unlike high mortality registered in the methanol 50 mg treatments. In both treatments at all doses applied, the end-point mortality was reached after 14 days. These findings are similar to (Liu et al., 2007) in which 33 Chinese herbs were evaluated for their toxicity against *S. zeamais* and *T. confusum*. Their data revealed that 14 out of the 33 plants tested had the hexane extracts causing mortality of *S. zeamais,* while methanol extracts of the same plants did not (Liu et al., 2007). In addition, feeding activity was observed in grains treated using 25 mg of both solvent extracts; while low feeding was observed in those treated with 50 mg of methanol extracts. Interestingly, no feeding activity was observed in grains treated with 50 mg hexane extracts. It is suggested that hexane extracts had more feeding deterrent bioactive constituents than methanol extracts. The results from the study agree with those conducted by Othira et al. (2009) using *H. Spicigera* solvent extracts. The study established that *H. Spicigera* hexane extract had higher feeding deterrence activity compared to either ethyl acetate or water extracts against the maize weevil (Othira et al., 2009).

High F_1 progeny numbers were realized in grains treated with 25 mg in both solvent extracts. This could be attributed to the ability of the weevils to bore oviposition pits while feeding. Grains treated with 50 mg of the solvent extracts had low or zero F_1 progeny values. It could be that at this dose the extracts seem to exert some antifeedant or ovicidal properties. Terpenoids isolated from some tropical rutales and sesquiterpene lactones from *Rhaponticum pulchrum* have been shown to have a phagorepressive effect against *S. oryzae* and *S. granaries* close relatives of *S. zeamais* (Cis et al., 2006; Omar et al., 2007).

Essential oils and powders of *H. spicigera* were found to exert the highest repellent action to the weevils, under the same testing conditions. Only the highest dose (100 mg) of hexane treated maize had marked repellency effect on the weevils, compared to the methanol solvent extracts (50 mg, 100 mg) and the lower (25 mg) hexane dose applied. From the results obtained, it is safe to suggest that the repellent action observed in the leaf powders is due to the action of essential oils from the leaves glandular trichomes, which rapture when disturbed to release the volatile oils. Likewise, Ogendo et al. (2004) found similar repellency against maize weevil using *Lantana camara* L. and *Tephrosia vogeli* Hook plant powders. Fractions containing oxygenated compounds of Alligator pepper and ginger have been shown in fourway olfactometer bioassays to account for the repellent activity against the maize weevil (Ukeh et al., 2009). The identified 3 major compounds in the behaviourally active fractions of *A. melegueta* and *Z. officinale* to be (S)-2-heptanol, (S)-2-heptyl acetate and (R)-linalool in a ratio of 1:6:3, and 1,8-cineole, neral and geranial in the ration of 5.48:1:2.13 respectively (Ukeh et al., 2009).

Chemical analyses of essential oils of *H. spicigera* leaves and inflorescence revealed that monoterpenes were the main compounds (>55%), with the main constituents being identified as α-pinene, β-phellandrene, α-thujene, β- pinene, limonene and terpinen-4-ol. The compositions of the oils were similar with respect to the major representative compounds. Terpenoids from Labiatae and Lauraceae have been shown to cause *S. oryzae* mortality (Rozman et al., 2007). The monoterpenes in *H. spicigera* are the most likely compounds involved in the suggested enzyme inhibition in *S. zeamais*. Generally, it is presumed that the crude essential oils from *H. spicigera* are not particularly dangerous to consumers since they have been used since antiquity in food preparations. Their potential toxic effects are also mitigated by the fact that they should easily evaporate when subjected to heat such as during cooking.

Locally available plant materials can be used to limit damage to stored produce caused by storage pests. In the present study, various plant extracts and powders of *H. spicigera* were tested for their efficacy at varied doses against the maize weevil (*S. zeamais*). All *H. spicigera* extracts tested showed some ability to protect maize against damage by *S. zeamais*, demonstrating a scientific rationale in grain protection by communities in Northern Uganda and Southern Sudan. Thus, it is highly unlikely that psychosomatic factors could have played a role in communities embracing this plant as a post-harvest grain protectant. The results indicate good potential for the use of as repellent and toxic agents in the management of the maize weevil. Further studies are however warranted to conclusively recommend its introduction in grain protection in the country.

CONCLUSION

Utilizing *H. spicigera* residues, as a pest management tool, in the protection of stored grains may be one of the more readily applicable uses of residue allelopathy. *Hyptis spicigera* extracts had various biological activities (fumigant, repellent and contact toxicity) against maize weevil, confirming the plants as a rich source of biochemicals that could be tapped for use as pesticides (Devkumar and Sukhdev, 1993). Powders, solvent extracts and essential oils showed repellent activity against the maize weevil, with essential oils being the most potent. Essential oils were shown to have potent fumigant activity against maize weevil. Additionally, solvent extracts exerted contact toxicity against maize weevil with anti feedant activity also being registered in hexane (50 mg) treatments. Modifying these *H. spicigera* extracts could lead to end products that are more active, selective or persistent. Among the well represented plant pesticides is "Pyrethrums" obtained from *Chrysanthemum cinarariaefolium* (Jacobson, 1976). If properly implemented in appropriate effective doses, *H. spicigera* plant extracts could save the resource poor farmer revenue which would have otherwise been lost through buying of synthetics and or storage pests losses. The results from the study demonstrated the scientific rationale for the use of *H. spicigera* in ethno-botanical grain storage practices.

REFERENCES

Abbot WS. (1925). A method of computing the effectiveness of an insecticide. J. Econ. Entomology, 18:265-267.

Aguirre C, Valde'z-Rodriguez S, Mendoza-Herna'ndes G, Dominguez RA, Blanco-Labra A. (2004). A novel 8.7KDa protease inhibitor from chan seeds (*Hyptis suaveolens*) inhibits proteases from he larger grain borer *Prostephanus truncatus* (Coleoptera:Bostrichidae). Comparative Biochemistry and Physiology. Part B: Biochemistry and Molecular Biology, 138(1):81-89.

Aycard JP, Kini F, Kam B, Gaydou M, Faure R. (1993). Isolation and identification of spicigera lactone: Complete ^{1}H and ^{13}C assignments using two dimensional NMR experiments. Journal of Natural Products, 56(7):1171-1173.

Bekele JA. (1995). Effects and use of some *Ocimum* plant species and their essential oils on some storage insect pests. PhD Thesis. Kenya.

Belmain RS, Neal EG, Roy DE, Golob P. (2001). Insecticidal and vertebrate toxicity associated with ethnobotanicals used as post-harvest protectants in Ghana. Food and Chemical Toxicology, 39(3):287-291.

Boakino MD, Connolly DJ, McLean S, Reynold FW, Tinto FW. (2003). α-pyrones and a 2(5H)-furanone from *Hyptis pectinata*. Phytochemistry, 64(7):1303-1307.

Boeke SJ, Baumgart RI, Van Loon, AJJ,Van Huis A, Dicke M, Kossou DK. (2004). Toxicity and repellence of African plants traditionally used for the protection of stored cowpea against *Callosobruchus maculatus*. Journal of Stored Products Research, 40(4):423-438.

Boeke SJ,Boersma MG, Alink MG, Joop JA, van Loon, Arnold van Huis, Dicke M, Rietjens IMCM. (2004). Safety evaluation of neem (*Azadirachta indica*) derived pesticides. Journal of Ethnopharmacology, 94(1):25-4.

Braga LS, CorrêaAS, Pereira EJG, Guedes, RNC. (2011). Face or flee? Fenitrothion resistance and behavioral response in populations of the maize weevil, *Sitophilus zeamais*. Journal of Stored Products Research, 47:161-167.

Buchanan GO, Williams L, Reese PB. (2000). Biotransformation of cadinane sesquiterpenes by *Beauveria bassiana* ATCC 7159. Phytochemistry, 54(1):39-45.

Sha SC, Liu QR, Liu ZL. (2010). Insecticidal activity and chemical composition of the essential oil of *Artemisia vestita* from China against *Sitophilus zeamais*. Biochemical Systematics and Ecology, 38:489–492.

Cis J, Nowak G, Kisiel W. (2006). Antifeedant properties and chemotaxonomic implications of sesquiterpene lactones and syringin from *Rhaponticum pulchrum*. Biochemiical Systematics and Ecology, 34(12):862-867.

Cosimi S, Rossi E, Cioni PL, Canale A. (2009). Bioactivity and qualitative analysis of s pests: evaluation of repellency against *Sitophilus zeamais* Motschulsky, *Cryptolestes ferrugineus* (Stephens) and *Tenebrio molitor* (L.). Journal of Stored Products Research, 45:125–132.

De Pury JMS. (1968). Crop pests of East Africa, Oxford University Press. 128-131.

Devkumar C, Sukhdev. (1993). In *Neem Research and Development* (eds Radhwa, N. S. and Parmar, B.S.), Society of Pesticide Science, New Delhi, pp. 63-96.

FAO. (1999). The use of spices and medicinals as bioactive protectants for grains. FAO Agricultural Services Bulletin No. 137. FAO, Rome.

FAO. (1998). Information Network on Post-Harvest Operations. www.fao.org

Finney DJ. (1978). Statistical methods in Biological assay 3rd Edn. Charles Griffin & Company, London, p508.

Fishel FM. Evaluation of Pesticides for Carcinogenic Potential. (2005). Florida Cooperative Extension Service, Institute of Food and Agricultural Sciences, University of Florida. PI-37 document. http://edis.ifas.ufl.edu.

Györi J, Varró P, Zielinska E, Banczerowski-Pelyhe I, Világi, I. (2007). Bensultap decreases neuronal excitability in molluscan and mammalian central nervous system. Toxicology in Vitro, 21(6):1050-1057.

Hassanali A, Lwande W, Ole-Sitayo N, Moreka L, Nokoe S, Chapya A. (1990). Weevil repellent constituents of *Ocimum suave* leaves and *Eugenia caryophyllata* cloves as grain protectants in parts of Eastern Africa. Discovery and Innovation, 2:91-95.

Ibrahim NA, Gamal BAEL. (2003). Effect of diazinon an organophosphate insecticide, on plasma lipid constituents in experimental animals. Journal of Biochemistry and Molecular Biology, 36:499-504.

Keita SM, Vincent C, Jean-Pierre S, Ramaswamy S, B'elanger A. (2000). Effects of various essential oils on *Callosobrunchus maculatus* (F.) (Coleoptera: Bruchidae). Journal of Stored Products Research, 36:355-364.

Kim Soon-II, Myung-Hee O, Hyung-Chang, C, Young-Joon A. (2003). Contact and fumigant activities of aromatic plant extracts against *Lasioderma serricorne* (Coleoptera: Anobiidae*)*. Journal of Stored Products Research, 39 (1):11-19.

Lecoeur S, Videmann B, Mazallon M. (2006). Effect of organophosphate pesticide diazinon on expression and activity of intestinal *P*-glycoprotein._Toxicology Letters, 161(3):200-209.

Liu LZ, Goh HS, Ho HS. (2007). Screening of Chinese medicinal herbs for bioactivity against *Sitophilus zeamais* Motschulsky and *Tribolium castaneum* (Herbst). Journal of Stored Products Research, 43(3):290-296.

Nguemtchouin MMG, M.B. Ngassoum, L.S.T. Ngamo, X. Gaudu, M. Cretin. (2010). Insecticidal formulation based on *Xylopia aethiopica* essential oil and kaolinite clay for maize protection. Crop Protection, 29:985-991.

Niber BT. (1994). The ability of powders and slurries from ten plant species to protect stored grain from attack by *Prostephanus truncatus* Horn. (Coleoptera: Bostrichidae) and *Sitophilus oryzae* L. (Coleoptera: Curculionidae). Journal of Stored Products Research, 30(4):297-301.

Obeng-Ofori D, Amitaye S. (2005). Efficacy of mixing vegetable oils with pirimiphos-methyl against the maize weevil, *Sitophilus zeamias* Motchulsky in stored maize. Journal of Stored Products Research, 41(1):57-66.

Ogendo JO, Deng AL., Belmain, SR, Walker DJ, Musandu AAO. (2004). Effects of Insecticidal Plant Materials, *Lantana camara* L. and *Tephrosia vogelli* Hook, on the quality Parameters of Stored Maize Grains. Journal of Food Technology in Africa, 9(1):29-35.

Ogendo JO. (2000). Current pest status of *Sitophilus zeamais* Motchulsky (Coleoptera:Curculionidae), control methods and constraints to safe on farm maize grain storage in Kasipul Division, Kenya. Progress report on Mini-survey conducted in Kasipul Division, Rachuonyo District, Kenya. 12th-22nd September, 1999. Msc. Thesis. Kenya.

Oliveira, EE, Guedes CRN, Tótola MR, De Marco Jr. P. (2007). Competition between insecticide-susceptible and -resistant populations of the maize weevil, *Sitophilus zeamais*. Chemosphere, 69(1):17-24.

Olwande J. (2009). Proceedings of a roundtable discussion on Kenya's food situation: challenges and opportunities. Held at Laico Regency Hotel, Nairobi on 18th September, 2009.

Omar S, Marcotte M, Fields P, Sanchez EP, Poveda L, Mata R, Jimenez A, Durst T, Zhang J, MacKinnon S, Leaman D, Arnason TJ, Philogène BJR. (2007). Antifeedant activities of terpenoids isolated from tropical Rutales. Journal of Stored Products Research, 43(1):92-96.

Onek LA, Othira JO, Birech R. (2003). Biological Control of *Striga hermonthica* and Common Grain Storage Insect Pests by Use of *Hyptis spicigera* and *Ocimum bacilicum. AICAD Research Mini- Reviews*, 1, 2003.

Othira JO, Onek LA, Deng LA, Omolo, E.O. (2009). *Insecticidal potency of* Hyptis spicigera *preparations against* Sitophilus zeamais (l.) *and* Tribolium castaneum *(herbst) on stored maize grains.* African Journal of Agricultural Research, 4(3):187-192.

Othira JO, Onek LA, Deng AL, Kemey J, Omolo EO. (2008). Potential application of *Hyptis spicigera* for biological control of *Striga hermonthica* infestation. African Journal of Agricultural Research, 3:747-752.

Pest Control Products Board (2004). List of registered pest control products in Kenya.

Ribiero MB., Guedes CNR, Oliviera EE, Santos JP. (2003). Insecticide resistance and synergism in Brazillian Populations of *Sitophilus zeamais* (Coleoptera: Curculionidae). Journal of Stored Products Research, 39(1):21-31.

Ristaino JB. (1998). Phaseout of Methyl Bromide Accelerated. The American Phytopathological Society online Feature-1998-0298.
Rogelio PM, Mabel FS, Carlos M. (2001).Application of molecular mechanics in the total stereochemical elucidation of spicigerolide, a cytotoxic 6-tetraacetyloxyheptenyl-5, 6-dihydro--pyrone from *Hyptis spicigera*. Tetrahedron, 57(1):47-53.
Rozman V, Kalinovic I, Korunic, Z. (2007). Toxicity of naturally occurring compounds of Lamiaceae and Lauraceae to three stored-product insects. Journal of Stored Products Research, 43(4):349-355.
Sanchez-Pena LL, Reyes BE, Lopez-Carillo L, Recio R, Moran-Martinez J, Cebrian ME, Quintanilla-Vega B. (2004). Organophosphorus pesticide exposure alters sperm chromatin structure in Mexican agricultural workers. Toxicology Applied Pharmacology, 196:108-113.
SAS Institute Inc. (2000). SAS/STAT® User's guide. Release 8.2 Edition. SAS Institute Inc., SAS Campus drive, Cary, NC 27513
Soon-II K, Young-Joon A, Do-Hyoung K, Han-Seung L. (2003). Insecticidal activities of aromatic plant extracts and essential oils against *Sitophilus oryzae* and *Callosobrunchus chinensis.* Journal of Stored Products Research, 39 (3):293-303.
Duangsamorn S, Fields PG, Chandrapatya A. (2011). Fumigant toxicity of essential oils from three Thai plants (Zingiberaceae) and their major compounds against *Sitophilus zeamais, Tribolium castaneum* and two parasitoids. Journal of Stored Products Research, 47:222-230.
Tapondjou AL, Alder C, Bouda H, Fontem AD, Reichmuth C. (2005). Bioactivities of cymol and essential oils of *Cupressus sempervirens* and *Eucalyptus saligna* against *Sitophilus zeamais* Motchulsky and *Tribolium confusum* du val. Journal of Stored Products Research, 41:91-102.
Tapondjou AL, Alder C, Bouda H, Fontem AD. (2002). Efficacy of powder and essential oil from *Chenoprodium ambrosioides* leaves as post-harvest grain protectants against six-stored product beetles. Journal of Stored Products Research, 38(4):395-402.
Ukeh DA, Birkett MA, Pickett JA, Bowmana AS, Luntz AJM. (2009). Repellent activity of alligator pepper, *Aframomum melegueta*, and ginger, *Zingiber officinale*, against the maize weevil, *Sitophilus zeamais*. Journal of Phytochemistry, 70:751–758.
Zoubiri S, Baaliouamer A. (2011). Chemical composition and insecticidal properties of some aromatic herbs essential oils from Algeria. Food Chemistry, 129:179-182.

15. Bioactivity of Mexican marigold (*Tagetes minuta*) essential oil against maize weevil (*Sitophilus zeamais*)

Ingrid Wekesa[1], Sylvia I.R. Murunga[1], Moses Makayoto[1], Gladys Mugo[1], Margaret Mulaa[2]
[1]Chemical Engineering Division, Kenya Industrial Research and Development Institute, Kenya.
[2]Kenya Agricultural Research Institute, P. O. Box 450 - 30200, Kitale, Kenya.
**Corresponding author: wngrid@yahoo.com, ingrid.wekesa@kirdi.go.ke*

SUMMARY

Maize is the most important staple food in Kenya. Regrettably, its production is threatened by post harvest losses due to agricultural pests which must be minimized. In the recent past, synthetic chemicals have been widely employed in controlling insect pests of maize. Though fast acting, synthetics offer ephemeral benefits and the shift in consumer demands has given the impetus for the search of sustainable 'green' storage pest control practices. Mexican marigold (*Tagetes sp*) is a weed species which has a strong pungent smell and has been used traditionally by communities, in West Pokot Kenya, as a biocontrol agent against storage insect pests. The weed is lined fresh in stores and is claimed to exert toxic/repellant cereal protectant activities against maize weevil, *Sitophilus zeamais* Motschulsky, which has been identified as the major insect pest of stored maize in Kenya. The objective of this study was to assess the toxic activity of Mexican marigold essential oil extracts against *Sitophilus zeamais,* a major insect pest of maize, in an effort to justify its use as an ethnobased grain protectant stored maize grains. The essential oils were obtained through hydro distillation and tested against the maize weevil in fumigation bioassays at dose rates of 20, 40, 80 and 100 µL. The treatments were carried out in a completely randomized design, replicated at least four times and results obtained as percentage counts. One way analysis of variance was used in analysis of data and results presented using means and standard errors. Means separation was achieved using Duncans Multiple Range Test. Results obtained showed that essential oils had strong fumigant toxicity against *S. zeamais*. The toxicity was majorly concentration dependent with total mortality (100%) being registered on the third day for the 80 µL and 100 µL dose rate. No significant differences were observed in mortality within all treatments at 6 days of exposure to the essential oils. The results agree with ethno-knowledge on grain protection practices among communities in West Pokot, Kenya, and suggest that Mexican marigold may be exploited for the development of biopesticides to be used in maize weevil management.

INTRODUCTION

Kenya imports approximately 7,000 tons of synthetic pesticides annually, valued at KShs. 4 billion (US$ 50 million). These pesticides are an assortment of insecticides, fungicides, herbicides fumigants, rodenticides, growth regulators, defoliators, proteins, surfactants and wetting agents. Of the total pesticide imports, insecticides account for about 40% in terms of volume (2,900 tons) and 50% of the total cost of pesticide imports (Birech et al., 2006). Incidentally, Kenya is the leading producer of a natural pesticide, pyrethrin, which is a broad-spectrum insecticide made from dried flowers of pyrethrum (*Chrysanthemum cinerariaefolium*). Up to 8,000 tons of dried flowers are produced annually. About 95% of the crude pyrethrin is exported to developed countries in the west; 60% to USA and 35% to Europe. Only 5% is used in Africa - Egypt and South Africa take 2% each and only 1% remains in Kenya (Birech et al., 2006). To note are associated drawbacks of synthetic insecticides, which include abuse and misuse, a common occurrence in Africa and, although the use of pesticides in Africa represents a small fraction of the global total, misuse is disproportionately high (PCPB, 2004).

The advent of synthetic insecticides in the protection of stored products and their associated drawbacks has refocused attention to ancient practices used in post harvest protection. Plants constitute a rich source of bioactive chemicals and may provide alternatives to currently used insect control agents (Kim et al., 2003). They are attractive candidates for use in integrated pest management since they are active against a limited number of species, including specific target pests and are often biodegradable to non- toxic products (Soon-Il et al., 2003). There are over 30 botanicals which have been identified in North Rift Kenya for pest control (Mulaa, 1998, 1999). However, their mode of action has not been established. Literature sources cite the use of Mexican marigold (*Tagetes minuta*) and hot pepper for control of aphids and African bollworm in tomatoes (Mulaa, 1998; 1999, 2001) and hot pepper powder and ash for maize stalk borer control (Mulaa, 1998) as examples of successes in the use of botanical pesticides. Identification of active ingredients in some promising botanical extracts for pest control indicated that the bioactive principles in *Tephrosia vogelii* seeds are 5-methoxy-7, 8 (2”,2”-dimethylpyran) flavanone or obovatin C methyl ether, tephrosin and deguelin. While the first one is a flavanone, the last two are rotenoids. Rotenoids and flavanones are established insecticidals and generally bioactive principles. The presence of rotenoids (and maybe the modified flavanones as well) in the plant, especially in the seed explains the insect control effect that extracts of *T. vogelii* have shown during field experiments conducted in Kitale (Midiwo and Mulaa, 2004).

Tagetes minuta (Mexican marigold) is an aromatic plant belonging to the Asteraceae family. The plant is touted as being rich in natural products chemistry and essential oils, which have found application in cosmetic, perfumery, food and beverage and medicine industries (Vasudevan et al., 1997). It has been established that the plant possesses potent nematicidal, antimicrobial and insecticidal activities against aphids (Tereschuck et al., 1997; Tomova et al., 2005). Studies by Batish et al. (2006) demonstrated its potential as a natural herbicide in the management of rice weeds when used in dried powder form. The plant is used by farmers in North Rift Kenya as a grain protectant against insect attack while in storage. The weed is lined fresh in stores and is claimed to exert toxic/repellant cereal protectant activities against maize weevil. Despite the cosmopolitan use of plants as grain protectants, knowledge of methods and efficacy of these plants is scanty. Most of the information is derived from anecdotal sources, which provide minimum information (FAO, 1998). The objective of this study was to investigate the effect of essential oils of Mexican marigold on maize weevil, *Sitophilus zeamais*.

MATERIALS AND METHODS

Plant collection and pretreatment

Mexican marigold plants, in flowering stage, were collected from Kitale. Pretreatment of the collected samples involved drying them under shade for five days. The dried plants were subjected to hydrodistillation to obtain essential oils which were collected over hexane. The essential oils were concentrated by blowing off using a stream of nitrogen and stored in glass vials away from light awaiting use in fumigation tests.

Sourcing and rearing of insects

This was done according to Wekesa et al. (2011). Weevils were reared on maize free from insecticides, isolated from the infested stock after a 7-day egg laying period, and kept at 17-28ºC, relative humidity 38-69% to obtain newly emerged weevils of the same population. These were then used for laboratory tests.

Fumigant toxicity of *T. minuta* essential oils

Ten pairs of 3 week old mixed sex weevils were placed in glass vials and capped. To the caps of the vials, filter papers were glued onto which the essential oils of *T. minuta* were introduced at rates of 10, 15, 20, 40, 80 and 100 μl. Controls contained similar set up but without essential oil. The treatments in the assays were laid out in a completely randomized design and the procedure was replicated at least 8 times. Insect mortality counts were made after 1, 2, 3, 6 and 7 days after treatment. The insects were considered dead if no movement was observed when placed under a light source.

Repellant activity of *T. minuta* essential oils

Repellent effects of the essential oil against the maize weevil were performed using a choice bioassay system according to Bekele (1995). Fifty mixed sex adult insects were used in the study, with essential oil treatments being incorporated at 100, 80, 40 and 20 μL. The treatments in the bioassays were laid out in a completely randomized design and replicated four times. The number of insects present in the control jar (N_C) and the treated jars (N_T) were recorded after 24 hours exposure. Percentage repellency (PR) values were computed using the method by Hassanali et al. (1990): $PR = [N_C - N_T / (N_C + N_T)] \times 100$.

RESULTS AND DISCUSSION

Fumigation effect of *T. minuta* essential oils

The preliminary results obtained indicate that *T. minuta* essential oils possess potent fumigant activity against *S. zeamais*. Total mortality was registered on the third day for the 80 µL and 100 µL treatments. Further decrease in dose rate pointed towards the toxicity being concentration dependent as observed for 1 to 3 days of essential oil exposure for all treatments (Figure 1). Significant ($P \leq 0.05$) differences from the control were observed for the first three days of exposure to the essential oils for the 40 µL treatment and on day 2 and 3 for the 20 µL treatment. Treatment at 15 and 10 µL essential oil were not significantly different ($P > 0.05$) from the control after two days of exposure. However, significant ($P \leq 0.05$) levels were realized for the 15 µL treatment on day 3, and day 6 and 7 for both (15 and 10 µL) treatments. The results obtained also showed over 50% mortality being achieved on day 1, 2 for 40 and 20 µL and day 3 for 15 and 10 µL treatments, respectively.

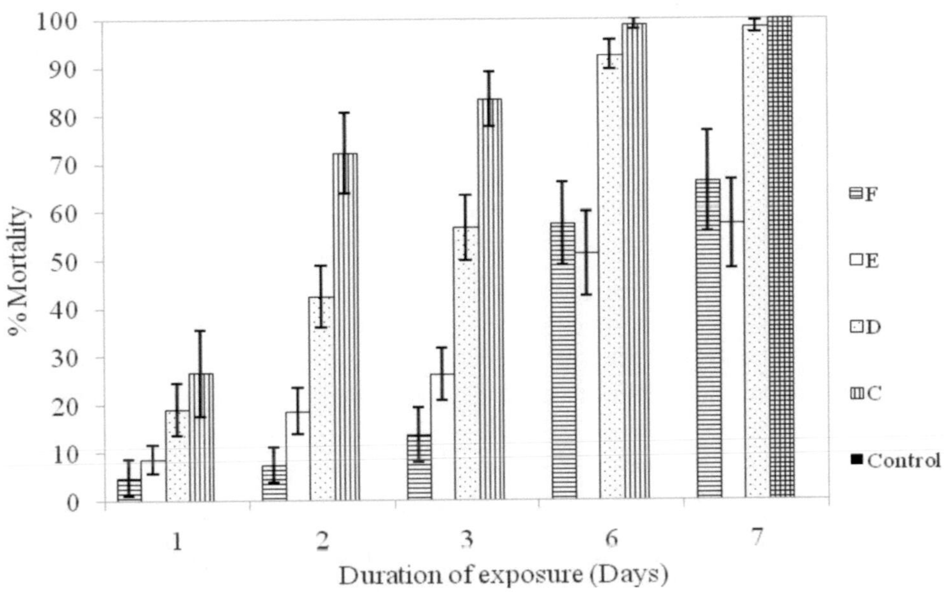

Figure 1. Fumigant effect of *T. minuta* essential oils on *S. zeamais*. C = 40µL, D = 20µL, E = 15 µL, F = 10µL

Mexican marigold thus joins other weed species whose insecticidal potential has been demonstrated. Some weeds species evaluated in laboratory studies for the ability to protect grains against insect attack are *Ageratum conyzoides*, *Lantana camara* and *Chromolaena odorata*. Their essential oils from leaves have been shown to cause significant mortality of *S. zeamais* in studies conducted by Bouda et al. (2001). In this study, mortality was found to increase with the concentration of essential oils and duration of exposure of the weevils to the treated substances. Powders of *Ocimum americanum* have been shown to dose dependently increase the mortality and F1 emergence of *P. truncatus*, *S. zeamais* and *Callosobruchus maculatus*, when included in 100 g maize at 0.5, 1.0 and 5% (Belmain et al., 2001). *Chemopodium ambrosiodes* powders and essential oils from leaves were evaluated for their efficacy as post-harvest grain protectants in laboratory studies conducted by Tapondjou et al. (2002). The investigations revealed that a dosage of 6.4% w/w (powders) induced total mortality of *S. zeamais* and *S. granarius* 48 hours after treatment. In addition, treatment with 0.2 μl/cm^2 essential oil killed 80-100% of the beetles (*Callosobruchus chinenesis, Acanthoscelides obtectus, S. granarius,* and *P. truncatus*) except *C. maculatus* and *S. zeamais* where this dosage induced only 20% and 5% mortality respectively (Tapondjou et al., 2002).

Repellent assays with *T. minuta* essential oil

Table 1 shows results obtained from repellency tests evaluating the potency of *T. minuta* essential oil treatments, against the maize weevil, in a choice bioassay system. The hydrodistillated essential oils of *T. minuta* showed significant ($P \leq 0.05$) repellency to *S. zeamais* weevils in the choice of test arm after 24 hour exposure (Table 1). *T. minuta* essential oils were repellent to maize weevil at all concentrations (100, 80, 40 and 20 μL) tested, with the highest concentration (100 μL) being significantly ($P \leq 0.05$) different from the others. The repellent action appeared to be dose dependent with more than 50% repellency being achieved for all dose rates assayed after 24 hour exposure.

Table 1. Repellent effect of *T. minuta* essential oils on *S. zeamais*

Treatment (μL)	% Mortality count
100	82.74^{b}
80	62.00^{a}
40	59.20^{a}
20	50.24^{a}

*Values with the same superscript letter in the column have no significant differences at $P \leq 0.05$.

The susceptibility of maize weevils to repellent action has also been demonstrated for the Asteraceae family by other researchers. Investigations by Nerio et al. (2009) using area preference assays showed that essential oils of *Tagetes lucida* (Asteraceae family), *Lippia origanoides* and *Eucalyptus citriodora* elicited repellent responses when tested at doses between 0.063 and 0.503 ml/cm^2. Studies by Ukeh et al. (2009) established Alligator pepper (*Aframomum melegueta*) and ginger (*Zingiber officinale*) vacuum distillates as having repellent action against maize weevil with the repellent activity being attributed to the fractions containing oxygenated compounds. Repellent potency was also established for bergamot (*Citrus bergamia*), fennel (*Foeniculum vulgare*) and lavandin (*Lavandula hybrida*) essential oils in filter paper and cup bioassay laboratory tests against *Sitophilus zeamais* by Cosimi et al. (2008).

CONCLUSION

The results suggest there is a scientific rationale to ethnobotanical claims on the ability of Mexican marigold to protect stored grain against insect attack. Furthermore, further studies are required to examine the behavioral responses of sexed adults and stability evaluation of the essential oils of *T. minuta*.

ACKNOWLEDGEMENTS

Thanks to Kenya Industrial Research and Development Institute (KIRDI) for funding the project and Kenya Agricultural Research Institute (KARI) for valuable assistance during the project period. We duly acknowledge the contribution of Mrs. Veronica Lokicho Menegich of Kasongwor village, Kipkomo Location, West Pokot and Dajopen Project, Kitale.

REFERENCES

Batish DR, Komal A, Singh HP, Kohli RK. (2007). Potential utilization of dried powder of *Tagetes minuta* as a natural herbicide for managing rice weeds. Crop Protection 26:566–571.

Belmain RS, Neal EG, Roy DE, Golob P. (2001). Insecticidal and vertebrate toxicity associated with ethnobotanicals used as post-harvest protectants in Ghana. Food and Chemical Toxicology, 39:287-291.

Rhoda B, Freyer B, Macharia J. (2006). Towards reducing synthetic pesticide imports in Favour of Locally Available Botanicals in Kenya. Conference on International Agricultural Research for Development, Bonn, October 11-13, 2006.

Bouda H, Tapondjou LA, Fontem AD, Gumedzoe DYM. (2001). Effects of essential oils from leaves of *Ageratum conyzoides*, *Lantana camara* and *Chromolaena odorata* on the mortality of *Sitophilus zeamais* (Coleoptera:Curculionidae). Journal of Stored Products Research, 37:103-109.

Cosimi, S, Rossi E, Cioni PL, Canale A. (2009). Bioactivity and qualitative analysis of some essential oils from Mediterranean plants against stored-product pests: Evaluation of repellency against Sitophilus zeamais Motschulsky, *Cryptolestes ferrugineus* (Stephens) and *Tenebrio molitor* (L.). Journal of Stored Products Research 45:125–132.

FAO. (1998). Information Network on Post-Harvest Operations. www.fao.org development in selected plant species. South African Journal of Plant and Soil, 15:1-5.

Hassanali A, Lwande W, Ole-Sitayo N, Moreka L, Nokoe S, Chapya A. (1990).Weevil repellent constituents of *Ocimum suave* leaves and *Eugenia caryophyllata* cloves as grain protectants in parts of Eastern Africa. Discovery and Innovation 2:91-95.

Kim Soon-II, Myung-Hee O, Hyung-Chang C, Young-Joon A. (2003). Contact and fumigant activities of aromatic plant extracts against *Lasioderma serricorne* (Coleoptera: Anobiidae). Journal of Stored Products Research, 39:11-19.

Midiwo J, Mulaa M. (2004). Preliminary report on Identification of active ingredients in Tephrosia vogelli (Unpublished).

Mulaa MA, Chesoli C, Wanyama J, Powon M. (1998). Gender Analysis in Participatory Evaluation of Tomato varieties and pest control options by farmers & Research Groups in Trans Nzoia district, Kenya. Edited by John Curry Margo Kooijman and Helga Recke. Institutionalising Gender in Agricultural Research: Experiences from Kenya. Kenya Agricultural Research Institute. Pp 125-139. ISBN No. 9966–879–23–4.

Mulaa M.A., Chesoli C., Wanyama J. and Powon M. (1999). Sustainable pest and disease control in tomato through use of indigenous technical knowledge and pesticides based on economic thresholds. Proceedings of KARI/DFID end of project conference March 23-27 pg 340-342.

Mulaa MA, Chesoli C, Kalama P, Osore P, Mugita A. (1999). Increasing farmer options for stalk borer control. A comparison between efficacy of natural products and artificial pesticides. Proceedings of KARI/DFID end of project conference March 23-27 pp 254-257.

Mulaa MA, Mugita A, Mose L, Hall D. (2001). Identification, Documentation and screening of Botanicals used by Farmers for crop pest control in North Rift Kenya. Proceedings of Conference on promotion of medicinal and other under utilized plant species for socio-economic development, Kenya Wildlife Service Training Institute Naivasha, Kenya 29th October-3rd November 2001.

Nerio LS, Olivero-Verbel J, Stashenko EE. (2009). Repellent activity of essential oils from seven aromatic plants grown in Colombia against *Sitophilus zeamais* Motschulsky (Coleoptera). Journal of Stored Products Research 45, 212–214

Pest Control Products Board. (2004). List of registered pest control products in Kenya.

Soon-II K, Young-Joon A, Do-Hyoung K, Han-Seung L. (2003). Insecticidal activities of aromatic plant extracts and essential oils against *Sitophilus oryzae* and *Callosobrunchus chinensis*. Journal of Stored Products Research, 39:293-303.

Tapondjou AL, Alder C, Bouda H, Fontem AD. (2002). Efficacy of powder and essential oil from *Chenoprodium ambrosioides* leaves as post-harvest grain protectants against six-stored product beetles. Journal of Stored Products Research, 38:395-402.

Tereschuk ML, Riera MV, Castro GR, Abdala LR. (1997). Antimicrobial activity of flavonoids from leaves of *Tagetes minuta*. J. Ethnopharm. 56:227–232.

Tomova BS, Waterhouse JS, Doberski J. (2005). The effect of fractionated *Tagetes* oil volatiles on aphid production. Entomol. Exp. Appl. 115, 153–159.

Vasudevan P, Kashyap S, Sharma S. (1997). *Tagetes*: a multipurpose plant. Biosci. Technol. 62, 29–35.

Wekesa I, Onek LA, Deng AL, Hasanali A, Othira JO. (2011).Toxicity and repellant potency of *Hyptis spicigera* extracts on *Sitophilus zeamais* motschulsky (Coleoptera: Curculionidae).

16. Pasting properties and *in vitro* digestibility of maize-based porridge flour blended with whey

Prudence K. Njiru[1*], Violet Mugalavai[2], Eunice A. Mewa[3], Anne W. Mutahi[3] and Calvin Onyango[1]
[1]Food Technology Division, Kenya Industrial Research and Development Institute, Nairobi, Kenya.
[2]Department of Family and Consumer Science, University of Eldoret, Eldoret, Kenya.
[3]Department of Food Science, Nutrition and Technology, University of Nairobi, Nairobi, Kenya.
**Corresponding author: prudence178@gmail.com*

SUMMARY

Whey is a collective term that refers to the serum or liquid part of milk that remains dissolved in the aqueous portion after coagulation of casein into curd during manufacture of cheese. Whey contain 55% of the total milk nutrients including soluble proteins, lactose, vitamins, calcium and phosphorus, and thus it can be a nutritious and cheap ingredient for the formulation of foods. The aim of this study was to determine the possibility of using whey and maize flour to develop a nutritious and low-dietary-bulk infant food. Porridge flour was prepared from native, malted or fermented maize flour with different levels of whey (0-40%). Proximate analysis of raw materials (without whey addition) revealed that native maize flour had higher crude fat (4.69 g/100 g), crude fiber (4.42 g/100 g), total ash (1.31 g/100 g), carbohydrate (85.19 g/100 g) and starch content than malted or fermented maize flour. Malted flour had higher reducing sugars (246.6 g/100 g), free amino nitrogen (40.01 g/100 g) damaged starch (6.33 g/100 g), crude protein (8.52 g/100 g) and diastatic power content (9.45°). Fermented flour had highest total titritable acidity (1.04 %) and the lowest pH (3.51). There was no significant difference ($P > 0.05$) in total soluble sugars for all the flours. The interaction effect of flour type (native, malted and fermented maize flour) and whey concentration was significant ($P \leq 0.001$) for consistency of cooked porridge, water solubility index, water absorption index and *in vitro* protein and starch digestibility. Pasting temperature, time to peak viscosity, breakdown viscosity and setback viscosity were significantly affected ($P \leq 0.001$) by the interaction effect of flour type (native, malted and fermented maize flour) and whey concentration. Flour type significantly ($P \leq 0.001$) affected peak viscosity and final viscosity. We concluded that (i) malting is beneficial in weaning food formulation because malt enzymes digest the grains resulting in dietary bulk reduction and improved digestibility; and (ii) the nutrient quality of malt porridge can be improved by adding whey.

INTRODUCTION

Whey is the yellow-green liquid that separates from the curd during the manufacture of cheese and casein (Smithers et al., 1996). It is generally considered a waste by-product of the dairy industry. Despite containing 55% of the total milk nutrients, including lactose, proteins, lipids, mineral salts and vitamins (Panesar and Kennedy, 2011), utilization of whey as human food is limited because of its disagreeable flavour (Branger et al., 1999). Furthermore, the high oxygen demand for biological degradation of whey solids makes its disposal very costly (Smithers et al., 1996; Marwaha and Kennedy, 1988). It is for this latter reason that about 50% of the total world cheese-whey production is treated and transformed into various food products and food ingredients (Marwaha and Kennedy, 1988). Whey can be added to non-alcoholic beverages (Sinha et al., 2007; Branger et al., 1999, Vojnovic et al., 1993), bread (Yousif et al., 1988) or biscuits (Sharma and Khetarpaul, 1997). It can also be incorporated in animal feed formulations or used as a microbial substrate to obtain various products such as ethanol, single-cell protein, enzymes, organic acids and biogas (Panesar and Kennedy, 2011; Siso, 1996). Ultrafiltration or microfiltration of liquid whey produces whey protein concentrate (protein concentrations less than 90%) or whey protein isolate (contains a minimum of 90% protein), which are valuable ingredients in the food industry.

Majority of people in Kenya depend on cereal grains as their staple food because of their limited income and high prices of animal foods. Cereals, such as maize, offer several challenges from a nutrition point of view, especially due to their limited quantity and amino acid profile of their protein fraction. The bioavailability of their minerals is limited due to relatively low mineral levels and the presence of phytic acid (Nout, 2009; Fageer et al., 2004). Consequently, several techniques, including genetic manipulation, processing (Sokrab et al., 2012) and fortication (Annan et al., 2005), have been used to improve biological utilisation of nutrients in maize. Simple technologies that are applied to mitigate these problems include fermentation or malting. Fermentation enhances flavour and texture; extends the shelf-life, enriches the nutrient quality; decreases the cooking time; and improves digestibility (Nout, 2009; Steinkraus, 2002). Malting improves protein quality and increases protein quantity (as a result of carbohydrate respiration), improves carbohydrate digestibility and palatability, and increases the energy and nutrient density (Belton and Taylor, 2004). Enrichment of commonly consumed staples such as porridge flour with dairy by-products, such as whey, is likely to be an effective strategy of further improving the nutrient quality. The objective of this study is to evaluate the influence of whey on the physicochemical and *in vitro* starch and protein digestibility of native, fermented and malted maize porridge flours.

MATERIALS AND METHODS

Preparation of native, malted and fermented maize flour

Maize (*Zea mays*) was purchased from the National Cereals and Produce Board, Nairobi, Kenya. The grains were cleaned and then washed in cold running tap water and sun-dried to a moisture content of about 10%. The grains were divided into three lots. Native flour was prepared by milling the grains using a disc mill having a 800 μm sieve. Malt flour was prepared by first steeping the grains for 24 hours in tap water at 24°C. The steep water was changed after 4 and 8 hours. After steeping, excess water was drained and the grains placed in perforated nylon bags and allowed to germinate for 72 hours at 24°C. Twice daily the bag was immersed for 10 min in tap water after which it was gently turned to avoid meshing of the roots and shoots. At the end of germination, the malt was sun-dried for 48 hours. The dried malt (together with external roots and shoots) was milled using a disc mill. Fermented flour was prepared by mixing milled flour with tap water in the ratio of 1:2. The contents were stirred, covered with aluminium foil and allowed to ferment spontaneously at 25°C for 48 hours. After fermentation, the slurry was spread thinly on aluminium trays and dried in a forced-air cabinet oven at 60°C for 24 hours. The fermented dried cake was milled using a disc mill and screened through a 800 μm sieve.

Composition of flours

Moisture, protein (N x 6.25), fiber, ash, and fat contents of the flours were determined according to AAAC Method 44-15A, 46-11A, 32-10, 08-01, and 32-25 (AACC, 2000), respectively. Total carbohydrate was determined as the difference between 100 and the sum of values for moisture, fat, protein, fibre and ash. Starch content was determined using the Megazyme Total Starch Assay Kit (Megazyme International Ireland Ltd., Co. Wicklow, Ireland). Total soluble sugars were determined by the phenol-sulphuric acid method (Dubois et al., 1956). Reducing sugars were determined using the Nelson-Somogyi alkaline copper reduction method (Somogyi, 1952). Damaged starch content was determined using the Megazyme Starch Damage Assay kit, K-SDAM 05/2008, (Megazyme International Ireland Ltd., Co. Wicklow, Ireland). Diastatic power (i.e. joint α- and β-amylase activity) was determined according to the procedure of American Society of Brewing Chemists (ASBC, 1958). Free amino nitrogen was determined by adding 1 g sample to 40 ml 5% trichloroacetic acid at 30°C and extraction carried out for 1 h at 30°C. The tubes were swirled, at 15 min intervals, to suspend the contents. The extract (10 ml) was centrifuged at 4,500 x *g* for 10 min and 1 ml of clear supernatant diluted to 25 ml with distilled water. The samples were subjected to ninhydrin assay according to AOAC Method 10.180 (AOAC, 1980). In order to determine pH, the sample (10 g) was mixed with 20 ml distilled water, homogenized to achieve uniformity and the pH measured using using an TOA pH Meter (HM–7B, Tokyo, Japan). The sample (10 ml) was pipetted into a conical flask and 2 drops of 1% phenolphthalein indicator added. Titration was done using 0.1 mol/l sodium hydroxide to a faint pink colour that had to persist for at least 30 seconds. The titer volume was noted and used to calculate the total titratable acidity (TTA), which was expressed as percentage lactic acid (i.e. % lactic acid = $A \times 0.009 \times 100/V$; where: A = ml of 0.1 mol/l sodium hydroxide required for titration; and V = ml of sample taken for the test, 0.009= is a constant. The experiments were designed as single factor single completely randomized design with three replicates. The results were subjected to one-way analysis of variance and differences in treatment means identified at $P \leq 0.05$ by Duncan's Multiple Range Test using GenStat Edition 13 software (VSN International Ltd, UK).

Formulation of maize-whey porridge flour

Sweet whey (5.5^0 brix, 5.8 pH) was donated by New Kenya Cooperative Creameries. Native, malted and fermented maize flours were blended with whey at ratios of 100:0, 90:10, 80:20, 70:30 and 60:40 on weight-basis. The flour-whey mixtures were dried using a forced-air electric cabinet drier at 40°C, and then milled using a disc mill fitted with a 800 μm screen.

Physical properties of maize-whey porridge flour

Water absorption index (or capacity) was determined by weighing the sample (5 g) into pre-weighed 50 ml centrifuge tubes. Distilled water (25 ml) was added to each tube, the cap secured, and the tubes vigorously hand-shaken for 5 seconds to suspend the flour. The suspension was allowed to solvate and swell for 20 minutes, with intermittent shaking at 5, 10, 15 and 20 minutes. The samples were centrifuged at 1,000 x *g* for 15 minutes after which the supernatant was decanted, and pellet drained by touching with an adsorbent-paper towel. Water absorption index was determined as the difference between the weight of the drained tube and the weight of the original empty tube. Water solubility index was determined by adding starch (1 g) into distilled water (20 ml) in test tubes and heated in a water bath at 60°C for 30 minutes. At the end of heating, the mixture was centrifuged at 1,200 x *g* for 20 minutes, and 10 ml of the supernatant decanted and dried to constant weight at 110^0C. Water solubility index was expressed as the percent of dried solid weight based on weight of the sample.

Pasting properties of the flours were determined using a viscograph-E (Brabender GmbH and Co. KG, Duisburg, Germany). The temperature-time conditions were: heating from 30°C to 95°C at the rate of 1.5°C/minute; holding at 95°C for 10 minutes; cooling from 95°C to 50°C at the rate of 1.5°C/minute; and holding at 50°C for 10 minutes. The parameters measured were paste temperature (temperature at which the paste viscosity starts to increase), peak viscosity (maximum hot paste viscosity), time to peak viscosity, breakdown (peak viscosity minus trough viscosity), final viscosity (highest viscosity during the second holding period) and setback (final viscosity minus trough viscosity).

Cooked paste consistency of the porridge was determined by adding 100 g flour to 1 litre tap water. Part of the water (750 ml) was boiled and the remaining cold water was mixed with the flour to make slurry. The slurry was added to the boiling water and stirred for 10 minutes. The porridge was incubated at 30°C for 1 hour before measuring the consistency using an HDP/FE forward extrusion cell of a TA.XT.*plus* Texture Analyzer equipped with a 50 kg load cell (Stable Micro Systems, Godalming, UK). Prior to measurement, the porridge was manually stirred to guarantee homogeneity and 100 g loaded into the cell. Compression force was measured at the following conditions: pre-test speed 10 mm/second, test speed 1 mm/second, trigger force 50 g, post-test speed 10 mm/second, compression distance 20 mm, outlet diameter of extrusion cell 3 mm. The average force after reaching a plateau (at 10-18 s) was used as an indicator of cooked paste consistency.

In vitro protein and starch digestibility of maize-whey porridge flour

In vitro soluble protein digestibility was determined by adding 200 mg sample to a 100 ml Erlenmeyer flask containing 35 ml 0.1 mol/l sodium citrate tribasic dihydrate (pH 2.0) with pepsin (1.5 g pepsin/1, Sigma P-7012). The mixture was incubated for 2 hours in a shaking water bath at 37°C then centrifuged at 10,000 rpm for 15 minutes. The residue was washed in 10 ml 0.1 mol/l phosphate buffer (pH 7.0) and recentrifuged at 10,000 rpm for 15 minutes, then resuspended in 35 ml 0.1 mol/l phosphate buffer (pH 8.0) with pancreatin solution (1.5 g pancreatin/1, Sigma P-1750). The mixture was incubated in a shaking water bath at 37°C for 1 h. This step will be followed by centrifugation at 10,000 rpm for 15 minutes, washing the residue in 10 ml phosphate buffer (pH 7.0) and recentrifugation at 10,000 rpm for 15 minutes. The residue was collected on nitrogen-free filter paper and washed with 10 ml phosphate buffer (pH 7.0). The dried residue was analyzed for nitrogen by the Kjeldahl method. Residual protein was subtracted from total protein and the difference expressed as a percent of the total protein and reported as IVSP digestibility. Soluble nitrogen was determined by weighing 1 g sample into 50 ml centrifuge tube and dispersed in 20 ml distilled water. The dispersion was mechanically shaken for 1 hour and centrifuged at 10,000 rpm for 15 minutes before collecting the supernatant. The residue was resuspended and centrifuged twice in 10 ml distilled water. The combined supernatants were analyzed for soluble nitrogen by the Kjeldhal method. *In vitro* insoluble protein (IVISP) digestibility was calculated as: ((Insoluble protein - Residual protein) / Insoluble protein x 100)), where insoluble protein = total protein - soluble protein; residual protein = protein remaining after pepsin hydrolysis.

In vitro starch digestibility was determined by dissolving 5 mg sample in 1 ml 0.2 mol/l phosphate buffer (pH 6.9). Porcine pancreatic amylase (20 mg) was dissolved in 50 ml of the same buffer and 0.5 ml added to the sample suspension and incubated at 37°C for 2 hours. 3-5 dinitrosalicyclic acid (DNS, 1 ml) was added quickly and the mixture heated for 5 minutes in a boiling water-bath. After cooling, the solution was made to 25 ml with distilled water and filtered prior to measurement of absorbance in a 5 mm cuvette at 510 nm using a spectrophotometer. A blank for each sample was prepared by incubating the sample first and 3-5 DNS added before addition of the enzyme solution. A standard curve was prepared using solutions containing known concentrations of maltose monohydrate. The curve was used to calculate the concentration of starch digestion products in test solutions. The values were expressed as mg maltose/g starch.

Water absorption index, water solubility index, consistency, pasting properties and *in vitro* protein and starch digestibility of the formulated blends experiments were designed as 3x5 factorial design with three replicates. Analysis of variance was carried out and the P-value used to determine significance of interaction between variables. The results were also subjected to one-way analysis of variance and differences in treatment means identified at $P \leq 0.05$ by DMRT.

RESULTS AND DISCUSSION

Composition of native, malted and fermented maize flour

Table 1 shows the proximate composition of native, malted and fermented maize flours. Carbohydrate, fibre, ash and fat contents declined after fermentation or malting. Malted flour had significantly higher ($P \leq 0.05$) protein content (8.52 g/100 g) than native flour (7.22 g/100 g) but this was not significantly different ($P > 0.05$) from fermented flour (7.71 g/100 g). Grains undergo diverse biochemical changes during malting or fermentation leading to redistribution of the contents of various nutrients. Fat content declines during germination because it is partially utilized for the production of energy required to support the biochemical and physiological changes occurring during germination (Elmaki et al., 1999; Mtebe et al., 1993). Fat content also decreases after fermentation due to utilization of fat as energy source by the microbial flora. Fibre content declines during germination due to partial enzymatic solubilization of the fibre (Elmaki et al., 1999). The decrease in carbohydrate and increase in protein content after fermentation can be attributed to a decrease of carbon ratio in the total mass, resulting in redistribution of nutrient percentages. When microorganisms utilize carbohydrates as an energy source and produce carbon dioxide as a by-product, nitrogen in the fermented slurry is concentrated leading to an increase in the proportion of protein in the total mass increases (Onyango et al., 2004). The reduction of ash content after malting or fermentation can be attributed to leaching of the minerals during steeping prior to malting and during fermentation.

Table 1. Proximate composition of native, malted and fermented maize flours*.

Flour type	Carbohydrate	Crude protein	Crude fiber	Total ash	Crude fat
Native	85.19^{c}	7.22^{a}	4.42^{c}	1.31^{c}	4.69^{c}
Malt	81.66^{a}	8.52^{b}	3.57^{b}	1.14^{b}	3.00^{a}
Fermented	84.46^{b}	7.71^{ab}	2.72^{a}	0.85^{a}	3.83^{b}

*Values are given on dry matter basis (g /100 g).
Values followed by the same superscript letter in the same column are not significantly different at $P > 0.05$

Native flour had significantly higher starch content (77 g/100 g, $P \leq 0.05$) than fermented (62.59 g/100 g) or malted flour (71.30 g/100 g). This finding is consistent with the observation made for carbohydrate content, which also declined after fermentation and malting. Starch is the main reserve polysaccharide in cereals and the principle source of carbohydrates. During malting, starch is hydrolysed into simple sugars by amylases resulting in a decrease in total starch. Reduction in starch content in the fermented flour may be attributed to amylolytic action of microorganisms in the fermenting mixture. Starch hydrolysis leads to the formation of sugars. Total soluble sugars content did not change significantly after malting or fermentation (Table 2). However, reducing sugars content increased significantly ($P \leq 0.05$) from 72.9 mg/100 g in native flour to 246.6 mg/100 g in malted flour and 251.4 mg/100 g in fermented flour (Table 2).

Table 2. Composition of native, malted and fermented maize flour.

Flour type	Starch (g/100g)	TSS (g/100g)	RS (mg/100g)	DS (g/100g)	DP (°)	FAN (g/l)	TTA (g/100g)*	pH
Native	77.00[b]	4.51[a]	72.9[a]	1.93[a]	5.03[a]	15.82[a]	0.23[a]	6.10[c]
Malted	62.59[a]	5.47[a]	246.6[b]	6.33[c]	9.45[b]	40.01[c]	0.52[b]	5.66[b]
Fermented	71.30[a]	5.61[a]	251.4[b]	2.96[b]	7.44[ab]	26.16[b]	1.04[c]	3.51[a]

Values followed by the same superscript letter in the same column are not significantly different at P > 0.05.
Values are given on dry-matter-basis except total titritable acidity and pH.
TSS: Total soluble sugars; RS: reducing sugars; DS: damaged starch; DP: diastatic power; FAN: free-amino nitrogen; TTA: total titratable acidity.
*TTA was measured as g lactic acid/100 g.

Malted flour had significantly higher ($P \leq 0.05$) damaged starch content (6.33 g/100 g) than fermented (2.96 g/100 g) or native flours (1.93 g/100 g, Table 2). Because of the impact of milling on starch granules, it is inevitable that starch granules will be partially or completely damaged, and therefore milled flour always has some damaged starch. However, whereas damaged starch content in malted flour was 230% higher than in native flour, it was only 50% higher in fermented flour. Soaking the grains or flour during malting and fermentation, respectively, softens the starch granules and makes them more susceptible to damage during milling.

Malted flour had higher diastatic power (9.09^0) and free amino nitrogen content (40.01 mg/l) than fermented or native flours (Table 2). Malt quality is assessed primarily in terms of diastatic power and free amino nitrogen (Dewar et al., 1997). Diastatic power is a measure of the joint α-amylase and β-amylase activity, whereas free amino nitrogen is a measure of the products of proteolysis. During germination, diastase activity increases (i.e. α-amylase is synthesized and β-amylase is activated) and starch reserves are degraded to soluble sugars (Subramanian et al., 1992). Simultaneously, endogenous proteinase and peptidase enzymes breakdown protein reserves of the grain into amino acids and peptides that are collectively referred to as free amino nitrogen (Elmaki et al., 1999; Dewar et al., 1997).

Fermented flour had significantly ($P \leq 0.05$) lower pH (3.51) than native (pH 6.10) or malted flours (pH 5.66) (Table 2). Titratable acidity, which is a measure of the total non-volatile acid in the product, was significantly higher ($P \leq 0.05$) in the fermented flour (1.04%) than in native (0.23%) or malted flour (0.52%). The low pH and high titratable acidity in fermented flour suggest the presence and activity of lactic acid bacteria which are largely responsible for the production of lactic acid during spontaneous fermentation of maize-based foods (Afoakwa et al., 2004). The acidity recorded in malted flour could be a result of hydrolysis of complex organic molecules (Adeyemo et al., 1992). The three main sources of acids in cereals are lipids, phytin and phosphorous, which are hydrolyzed to fatty acids, acid phosphates and free amino acids.

Physical properties of maize-whey porridge flour

The interaction effect of flour type (native, malted and fermented maize flour) and whey concentration was significant ($P \leq 0.001$) for water absorption index (Table 4). Water absorption index was significantly higher ($P \leq 0.05$) in fermented flour than in native or malted flour at all levels of whey addition (Table 3). Water absorption index did not change ($P > 0.05$) with addition of whey in fermented flour but it tended to increase with increasing whey content in native or malted flour. High water absorption index reflects high starch content. Malted flour had lower water absorption index because enzymatic activity had disrupted most of starch granules and thus decreased the water binding capacity. Malting produces a complement of enzymes capable of converting cereal starch to simple sugars. During malting of grains, α- and β-amylase activity is developed. These enzymes effectively degrade starch granules reducing their water binding capacity and this is beneficial in weaning food formulation because it leads to dietary bulk reduction.

Table 3. Water absorption and solubility indices of maize-whey porridge flours.

Type of flour	Water absorption index				
	100:0*	90:10	80:20	70:30	60:40
Native	0.915^{ab}_{y}	0.920^{b}_{y}	0.915^{ab}_{x}	0.918^{ab}_{x}	0.939^{c}_{y}
Malted	0.909^{a}_{x}	0.909^{a}_{x}	0.917^{b}_{x}	0.919^{b}_{x}	0.922^{c}_{x}
Fermented	0.930^{a}_{z}	0.929^{a}_{z}	0.930^{a}_{y}	0.930^{a}_{y}	0.933^{a}_{y}
	Water solubility index				
Native	3.20^{a}_{y}	4.94^{b}_{y}	5.58^{bc}_{y}	3.75^{a}_{x}	6.25^{c}_{y}
Malted	4.52^{a}_{z}	7.70^{bc}_{z}	8.28^{c}_{z}	6.99^{b}_{z}	9.74^{d}_{z}
Fermented	1.91^{a}_{x}	2.14^{a}_{x}	2.74^{b}_{x}	5.78^{d}_{y}	3.64^{c}_{x}

*Ratio of maize flour to whey.

Values followed by the same superscript letter (a-c for water absorption index, a-d for water solubility index) in the same row are not significantly different at $P > 0.05$. Values followed by the same subscript letter (x-z) in the same column are not significantly different at $P > 0.05$.

The interaction effect of flour type and whey concentration was significant ($P \leq 0.001$) for water solubility index (Table 4). Water solubility index was lowest in flours without whey addition and highest in flours treated with 40% whey (Table 3). Irrespective of the whey content, malted flours had higher water solubility index than native or fermented flours. Starch solubility represented by water solubility index results from the leaching of amylose, which dissociates from and diffuses out of granules during swelling. This leaching represents a transition from order to disorder within the starch granules that occurs when starch containing water is heated (Tester and Morrison, 1990).

Table 4. F-values of water solubility index and water absorption index of maize-whey porridge flours.

Variable	Water absorption index	Water solubility index
Flour type	29.94***	500.61***
Whey concentration	7.12***	99.43***
Flour type x whey concentration	3.35**	35.53***
r^2	0.69	0.97

**significant at $P \leq 0.01$.
***significant at $P \leq 0.001$.
r^2: Coefficient of determination.

The pasting properties of native and fermented maize porridge flour enriched with whey are presented in Table 5. Malted flours did not show a response to the pasting profile because the hydrolytic activity of endogenous amylases, which are very stable at the pasting temperature regime, ensured continual starch hydrolysis with increasing temperature. Cereal amylases lose their activity at 70-80°C (Sluimer, 2007), which is within the range of the onset pasting temperature (72.95-78.95°C) of maize flour (Table 5). The temperature at which the first detectable increase in viscosity is measured is the pasting temperature of a sample. The interaction effect of flour type (native, malted or fermented maize flour) and whey concentration was significant ($P \leq 0.001$) for pasting temperature (Table 6). Pasting temperature gives an indication of the minimum temperature required to cook a given sample (Otegbayo et al., 2006). This can have implication on stability of other components in a formulation and also influence energy cost. It is essential that a product attain its pasting temperature early during processing to ensure swelling, gelatinization and subsequent gel formation. The difference in gelatinization temperature among different flours may be attributed to the difference in size, form and distribution of starch granules in the flours, and to the internal arrangement of starch fractions within the granule (Kaur and Singh, 2005).

Peak viscosity was significantly affected ($P \leq 0.001$) by the flour type only (Table 6). Peak viscosity is an indicator of the strength of paste formed during processing (Niba et al., 2001). It shows the viscosity of the cooked starch and reflects the ability of the starch to swell freely before their physical breakdown (Singh et al., 2003). These results confirm that fermentation does not reduce the viscosity of cereal slurries (Mensah et al., 1991). This means that the microbial flora in these slurries is unable to synthesize amylases that are required to catalyze breakdown of starch into monosaccharaides and oligosaccharides.

Setback viscosity is determined as difference between the final viscosity and trough viscosity. The interaction effect of flour type and whey concentration significantly affected ($P < 0.001$) setback viscosity (Table 6). Setback values represent the tendency of starch to gel. Native flour treated with 40% whey had the highest setback viscosity while fermented flour treated with 40% whey had the lowest. Setback viscosity gives an indication of the retrogradation tendency of starch. Foods with lower setback viscosities have higher potential for retrogradation than those having higher setback viscosities (Niba et al., 2001).

The interaction effect of flour type (native, malted or fermented maize flour) and whey concentration was significant ($P \leq 0.001$) for breakdown viscosity (Table 6). Breakdown viscosity is the difference between peak viscosity and trough viscosity. The breakdown value is an index of paste or gel stability during processing. Lower breakdown values indicate higher paste or gel stability during cooking. Cohesiveness of starch is attributed to the breakdown viscosities of starch molecules during heating and stirring. Starch may be predisposed to breakdown due to early gelatinization because it undergoes a longer period of shear (Moorthy et al., 1992).

Final viscosity was significantly affected ($P \leq 0.001$) by the flour type only (Table 6). Fermented flour had higher final viscosity than the native flour irrespective of the whey concentration (Table 5). During the final cycle of cooling (from 95 to 50°C), the viscosity increased owing to the alignment of the chains of amylose (Flores-Farias et al., 2000). Increase in final viscosity may be due to aggregation of the amylose molecules (Miles et al., 1985). Final viscosities of flour mixtures are important in determining the ability of a sample to form a gel during processing (Niba et al., 2001). It is an indication of whether the starch material forms a gel or a paste on cooling, and also indicates the strength of cooked paste.

The consistency of the cooked porridges blended with different whey concentrations was determined as the force (N) required to extrude the porridge through a HDP/FE forward extrusion cell attached to a TA.XT *plus* Texture Analyzer (Stable Micro Systems, Godalming, UK). The interaction effect of flour type (native, malted and fermented maize flour) and whey concentration significantly affected ($P \leq 0.001$) the consistency of cooked porridge (Table 6). Malted gruel was very "thin" and passed through the 3 mm die exit before the trigger force (50 g) was reached. This was attributed to the high diastatic activity of the flour (Table 7), which ensured that starch was degraded during cooking resulting in loss of water-binding capacity. This property is essential in the preparation of high energy density porridges, as it enables utilization of greater quantities of flour to produce porridge with the right consistency. Native or fermented flours had detectable forces because they lack significant enzymatic activity to ensure starch breakdown - hence the cooked native or fermented porridges were starchy and high in dietary bulk. This is because when cooked, the starch granules break open, absorb water and swell, making the gruel too thick to swallow. Native or fermented gruel without whey addition had significantly higher ($P \leq 0.05$) extrusion forces (7.88 N and 2.79 N respectively) than those blended with 10-40% whey (Table 7).

In vitro protein and starch digestibility of maize-porridge flour enriched with whey

Table 8 shows the *in vitro* insoluble and soluble protein and starch digestibility of native, malted or fermented porridge treated with whey. The interaction effect of flour type and whey concentration was significant ($P \leq 0.001$) for protein and starch digestibility (Table 9). Malted flour had significantly ($P \leq 0.05$) higher protein and starch digestibility than native or fermented flour at all levels of whey addition (Table 8). The biochemical modification in grains during malting is advantageous in producing malt with improved *in vitro* starch and protein digestibility. The increase in *in vitro* protein digestibility could be due to elimination of anti-nutrients (e.g. hydrolysis of phytic acid) thus rendering the proteins more susceptible to enzymatic hydrolysis. The improvement in starch digestibility is because malting hydrolyses into shorter chain dextrins and oligosaccharides that are more susceptible to digestion.

Table 5. Pasting properties of native (N) and fermented (F) maize flour treated with different whey concentrations.

Whey (%)	PT (^{0}C)		PV(BU)		T-PV (min)		BV (BU)*		FV (BU)		SV (BU)**	
	N	F	N	F	N	F	N	F	N	F	N	F
0	73.50^{ab}_{x}	80.30^{b}_{y}	439.5^{ab}_{x}	905.5^{a}_{y}	$40{:}21^{b}_{x}$	$44{:}05^{ab}_{y}$	81.5^{a}_{x}	122.5^{b}_{y}	959.5^{ab}_{x}	1822.0^{a}_{y}	578.5^{a}_{x}	1126.0^{c}_{y}
10	72.95^{a}_{x}	76.90^{a}_{y}	370.0^{a}_{x}	903.0^{a}_{y}	$39{:}18^{a}_{x}$	$43{:}30^{a}_{y}$	77.0^{a}_{x}	191.0^{d}_{y}	814.5^{a}_{x}	1796.0^{a}_{y}	491.5^{a}_{x}	1174.0^{c}_{y}
20	74.45^{b}_{x}	77.65^{a}_{y}	392.0^{a}_{x}	973.0^{a}_{y}	$40{:}24^{b}_{x}$	$44{:}02^{ab}_{y}$	128.5^{a}_{x}	160.5^{c}_{x}	902.0^{ab}_{x}	1854.0^{a}_{y}	652.0^{a}_{x}	1128.0^{c}_{y}
30	74.70^{b}_{x}	79.80^{b}_{y}	399.5^{a}_{x}	936.0^{a}_{y}	$40{:}46^{b}_{x}$	$44{:}14^{b}_{y}$	125.0^{a}_{x}	127.0^{b}_{x}	937.0^{ab}_{x}	1718.0^{a}_{y}	653.5^{a}_{x}	994.0^{b}_{x}
40	78.95^{c}_{x}	81.05^{b}_{y}	547.5^{b}_{x}	900.5^{a}_{y}	$43{:}03^{c}_{x}$	$44{:}31^{b}_{x}$	113.0^{a}_{x}	82.0^{a}_{x}	1211.5^{b}_{x}	1602.0^{a}_{x}	1004.0^{b}_{x}	869.0^{a}_{x}

Values followed by the same superscript letter (a-d) in the same column are not significantly different at $P > 0.05$.
Values followed by the same subscript letter (x and y) in the same row for each pasting property are not significantly different at $P > 0.05$.
*Breakdown viscosity = peak viscosity – trough viscosity.
** Setback viscosity = final viscosity – trough viscosity.
PT: Pasting temperature; PV: peak viscosity; T-PV: time to peak viscosity; BV: breakdown viscosity; FV: final viscosity; SV: setback viscosity.

Table 6. F values of consistency and pasting properties of native, malted and fermented maize flour treated with different whey concentrations.

Variable	Consistency (N)	PT (^{0}C)	PV (BU)	T-PV (min)	BV (BU)	FV (BU)	SV (BU)
Flour type	253.13***	295.75***	522.56***	914.91***	18.45**	130.25***	89.54***
Whey concentration	194.81***	47.20***	ns	53.99***	6.13**	ns	ns
Flour type x whey concentration	126.64***	10.78***	ns	21.88***	10.78***	ns	12.12***
r^2	0.98	0.96	0.97	0.98	0.80	0.87	0.88

*Significant at $P \leq 0.05$; **significant at $P \leq 0.01$; ***significant at $P \leq 0.001$.; ns not significant at $P > 0.05$.
r^2: Coefficient of determination.
PT: Pasting temperature; PV: peak viscosity; T-PV: time to peak viscosity; BV: breakdown viscosity; FV: final viscosity; SV: setback viscosity.

Table 7. Consistency (N) of cooked porridges treated with different whey concentrations.

Type of flour	Ratio of flour to whey				
	100:0	90:10	80:20	70:30	60:40
Native	7.88^{e}_{y}	2.87^{c}_{y}	1.66^{a}_{x}	2.47^{b}_{x}	3.36^{d}_{x}
Malted	nd*	nd	nd	nd	nd
Fermented	2.79^{c}_{x}	1.63^{a}_{x}	2.28^{b}_{y}	2.37^{b}_{x}	2.26^{b}_{y}

Values followed by the same superscript letter in the same row (a-e) are not significantly different at $P > 0.05$.
Values followed by the same subscript letter in the same column (x-z) are not significantly different at $P > 0.05$.
*nd: No force recorded because the gruel was very "thin" and passed through the cylinder with 3 mm die before the trigger force (50 g) was reached.

Table 8. *In vitro* insoluble and soluble protein digestibility (%) and *in vitro* starch digestibility of native, malted and fermented maize flour treated with different whey concentrations.

Type of flour	Ratio of flour to whey				
	100:0	90:10	80:20	70:30	60:40
	In vitro insoluble protein digestibility (%)				
Native	52.52^{ab}_{x}	48.67^{a}_{y}	51.49^{ab}_{x}	58.67^{b}_{y}	45.09^{a}_{x}
Malted	61.33^{a}_{y}	57.96^{a}_{z}	59.66^{a}_{y}	79.55^{c}_{z}	73.03^{b}_{y}
Fermented	51.90^{c}_{x}	32.27^{a}_{x}	49.62^{c}_{x}	45.06^{bc}_{x}	37.82^{ab}_{x}
	In vitro soluble protein digestibility (%)				
Native	54.67^{a}_{x}	51.70^{a}_{y}	54.87^{a}_{x}	61.42^{b}_{y}	48.66^{a}_{x}
Malted	66.55^{a}_{y}	64.25^{a}_{z}	65.89^{a}_{y}	82.42^{c}_{z}	75.65^{b}_{y}
Fermented	59.56^{c}_{xy}	43.87^{a}_{x}	57.07^{bc}_{x}	52.33^{b}_{x}	45.77^{a}_{x}
	In vitro starch digestibility (mg maltose/g starch)				
Native	79.9^{a}_{x}	164.5^{bc}_{x}	122.1^{ab}_{x}	206.5^{cd}_{x}	253.2^{d}_{x}
Malted	262.3^{a}_{y}	287.0^{a}_{y}	428.2^{bc}_{y}	365.6^{b}_{y}	468.6^{c}_{z}
Fermented	60.2^{a}_{x}	217.0^{b}_{xy}	108.8^{a}_{x}	210.4^{b}_{x}	328.2^{c}_{y}

Values followed by the same superscript letter (a-c) in the same row are not significantly different at P > 0.05.
Values followed by the same subscript letter (x-y) in the same column are not significantly different at P > 0.05.

Table 9. F-values of *in vitro* insoluble and soluble protein digestibility and starch digestibility of native, malted and fermented maize flour treated with different whey concentrations.

Variable	*In vitro* insoluble protein digestibility (%)	*In vitro* soluble protein digestibility (%)	*In vitro* starch digestibility
Flour type	139.80***	179.36***	174.22***
Whey concentration	17.70***	19.87***	54.13***
Flour type x whey concentration	9.88***	11.46***	7.47***
r^2	0.90	0.92	0.95

**significant at $P \leq 0.01$; ***significant at $P \leq 0.001$.
r^2: Coefficient of determination.

CONCLUSION

The aim of the study was to determine the possibility of using whey and maize flour to develop a nutritious and low-dietary-bulk infant food. Addition of whey improved the protein and starch digestibility and consistency of the porridge flours. It is recommended that native, malted or fermented maize flour should be blended with whey in order to improve the nutritional value.

REFERENCES

AACC International (2000). Approved methods of the American Association of Cereal Chemists. 10th Edition. St. Paul MN., USA.

Adeyemo SO, Oloyode OB, Odutuga AA. (1992). Biochemical analysis of germinating white maize (Zea mays). Nigeria Journal of Nutrition Science, 13:14-18.

Afoakwa EO, Sefa-Dedeh S, Sakyi-Dawson E. (2004). Effects of cowpea fortification, dehydration method and storage time on some quality characteristics of maize-based traditional weaning foods. African Journal of Food and Nutrition Development, 4(1): 1-17.

Annan NT, Plahar WA, Poll L, Jakobsen M. (2005) Effect of soybean fortification on Ghanaian fermented maize dough. International Journal of Food Sciences and Nutrition, 56(5): 315-/326.

AOAC. (1980). Approved Methods 10.180. Association of Official Analytical Chemists, Washington DC, USA.

ASBC (1958). Methods of Analysis of the American Society of Brewing Chemists. 6th Edition, St. Paul MN., USA.

Belton PS, Taylor JRN. (2004). Sorghum and millets: protein sources for Africa. Trends in Food Science and Technology, 15:94-98.

Branger EB, Sims C, Schmidt RH, O'Keefe SF, Cornell JA. (1999). Sensory characteristics of cheese whey and grape fruit blends and changes during processing. Journal of Food Science, 64:180-184.

Dewar J, Taylor JRN, Berjak P. (1997). Effect of germination conditions, with optimised steeping, on sorghum malt quality-with particular reference to free amino nitrogen. Journal of the Institute of Brewing, 103:171-175.

Elmaki HB, Babiker EE, El Tinay AH. (1999). Changes in chemical composition, grain malting, starch and tannin contents and protein digestibility during germination of sorghum cultivars. Food Chemistry, 64:331-336.

Fageer ASM, Babiker EE, El Tinay AH. (2004). Effect of malt pretreatment and / or cooking on phytate and essential amino acids contents and in vitro protein digestibility of corn flour. Food Chemistry, 88:261–265.

Flores-Farias R, Martinez-Bustos F, Salinas-Moreno Y, Chang YK, Hernandez JS, Rios E. (2000). Physicochemical and rheological characteristics of commercial nixtamalised Mexican corn flours for tortillas. Journal of the Science of Food and Agriculture, 80:657–664.

Kaur M, Singh N. (2005). Studies on functional, thermal and pasting properties of flours from different chickpea (Cicerarietinum L.) cultivars. Food Chemistry, 91(3): 403–411.

Marwaha SS, Kennedy JF. (1988). Review: whey pollution problem and potential utilization. International Journal of Food Science and Technology 23:323–336.

Moorthy SN, Rickard J, Blanshard JMV. (1992). Starch properties in relation to cooking quality of cassava. In Roca, W. M. and Thro, A. M. (Eds.), Proceedings of the First International Scientific Meeting, Cassava Biotechnology Network, Working document No.123. Cali, Colombia:CIAT.

Miles MJ, Morris VJ, Orford PD, Ring SG. (1985). The roles of amylose and amylopectin in the gelation and retrogradation of starch. Carbohydrate Research, 135:271–281.

Mtebe K, Ndabikunze BK, Bangu NTA, Mwemezi E. (1993).Effect of cereal germination on the energy density of togwa. International Journal of Food Sciences and Nutrition, 44:175-180.

Niba LL, Bokanga MM, Jackson FL, Schlimme DS, Li BW. (2002). Physicochemical properties and starch granular characteristics of flour from various Manihot esculenta (cassava) genotypes. Journal of Food Science, 67(5):1701-1705.

Nout MJR. (2009). Rich nutrition from the poorest – cereal fermentations in Africa and Asia. Food Microbiology, 26:685-692.

Onyango C, Noetzold H, Bley T, Henle T. (2004). Proximate composition and digestibility of fermented and extruded uji from maize-finger millet blend. Lebensmittel Wissenschaft und Technologie, 37:827-832.

Otegbayo BO, Aina JO, Asiedu R, Bokanga M. (2006). Pasting characteristics of fresh yam (Dioscoreaspp.) as indicator of food textural quality in a major yam food product-pounded yam. Food Chemistry, 99: 663-669.

Panesar PS, Kennedy JF. (2011). Biotechnological approaches for the value addition of whey. Critical Reviews in Biotechnology, 1-22.

Subramanian V, Murty DS, Rao NS, Jambunathan R. (1992). Chemical changes and diastatic activity in grains of sorghum (Sorghum bicolor) cultivars during germination. Journal of the Science of Food and Agriculture, 58: 35-40.

Sharma A, Khetarpaul N. (1997). Development of products incorporating fermented rice-dehulledblackgram-whey blends; their sensory and nutritional evaluation. Nahrung, 41 (1): 22-24.

Singh N, Singh J, Kaur L., Sodhi NS, Gill SB. (2003).Morphological, thermal and rheological properties of starches from different botanical sources. Food Chemistry, 81: 219-231

Sinha R, Radha C, Prakash J, Kaul P. (2007). Whey protein hydrolysate: Functional properties, nutritional quality and utilization in beverage formulation. Food Chemistry, 101: 1484–1491

Siso MIG. (1996). The biotechnological utilization of cheese whey: A review. Bioresource Technology, 57: 1-11.

Smithers GW, Ballard FJ, Copeland AD, De Silva KJ, Dionysius DA, Francis GL, Goddard C, Grieve PA, Mcintosh GH, Mitchell IR, Pearce RJ, Regester GO. (1996). New opportunities from the isolation and utilization of whey proteins. Journal of Dairy Science, 79, 1454-1459.

Sokrab AM, Ahmed IAM, Babiker EE. (2012). Effect of malting and fermentation on antinutrients, and total and extractable minerals of high and low phytate corn genotypes. International Journal of Food Science and Technology, 47, 1037–1043.

Steinkraus KH. (2002). Fermentations in world food processing. Comprehensive Reviews in Food Science and Food Safety, 1: 23-32.

Tester RF, Morrison WR. (1990). Swelling and gelatinization of cereal starches. I. Effects of amylopectin, amylose, and lipids. Cereal Chemistry, 67: 551-557.

Vojnovic V, Ritz M, Vahcicet H. (1993). Sensory evaluation of whey-based beverages. Die Nahrung, 37, (3), 246-251.

17. Development of anthracnose tolerant sorghum (*Sorghum bicolor* L. Moench) through genetic transformation with *chitinase* and *chitosanase* genes from *Trichoderma harzianum*

Linus Moses K'osambo-Ayoo[1*], Maram Bader[2] and Dirk Becker[2]
[1]Food Technology Division, Kenya Industrial Research and Development Institute, Nairobi, Kenya
[2]Developmental Biology and Biotechnology, Biocenter Klein Flottbek, University of Hamburg, Ohnhorststrasse 18, 22609 Hamburg, Germany.
[*]*Corresponding author e-mail: kosambosan@yahoo.com*

SUMMARY

Fungal diseases such as anthracnose which is caused by *Colletotrichum sublineolum* reduce sorghum yields. Genetic engineering can be used to confer tolerance to plant diseases such as anthracnose. The tolerance can be developed by introducing genes encoding antifungal proteins such as chitinases and chitosanases that hydrolyse fungal cell wall. Particle bombardment was used to genetically transform a sorghum genotype from Kenya, KAT 412, with *chitinase* (*harchit*) and *chitosanase* (*harcho*) genes isolated from *Trichoderma harzianum*. The genetic transformation also included introduction of herbicide resistance *pat* gene. Successful genetic transformation of KAT 412 with the two antifungal genes and the herbicide resistance gene was achieved. Transgenic sorghum plants, KOSA-1-3, coexpressing the two anti-fungal genes and the herbicide resistance *bar* gene were developed. Expression of the *harchit* and *harcho* in the transformants was confirmed by southern blot and quantitative real time PCR. *In planta* and *ex planta C. sublineolum* infection assays were carried out using one-week old seedlings to determine tolerance to anthracnose. Seedlings from a transgenic line, KOSA-1, were found to be significantly more tolerant to anthracnose than the parent wild type, KAT 412. The transgenic line was further compared with other wild type sorghum cultivars. The comparison revealed a genotype-dependent difference in anthracnose response. The transgenic line KOSA-1 was more tolerant than the SDSH 513 but less tolerant than KAT L5. This demonstrated the existence of genetic diversity, which together with the transgenes, can be utilised to pyramid genes for higher tolerance to anthracnose. The two antifungal genes introduced into sorghum genome can be introduced into other agronomic elite sorghum lines for fungal diseases resistance and enhanced productivity.

INTRODUCTION

Sorghum (*Sorghum bicolor* (L.) Moench) is one of the crops that are traditionally produced in marginal areas (Wenzel and Van Rooyen, 2001; Machado et al., 2002; Gebeyehu et al., 2004). It is a primary staple in the semiarid tropics of Africa and Asia for over 300 million people (ICRISAT, 2010). Sorghum is attacked by fungal, bacterial and viral pathogens causing root, stalk, foliar, panicle, and caryopsis diseases (Sutton, 1980; Waniska, 2001; Bueso et al., 2000; Prom et al., 2005). One of the fungal sorghum diseases of significance is anthracnose caused by *Colletotrichum sublineolum* (Sutton, 1980; Hammer et al., 1988; Nicholson and Epstein, 1991). Genetic engineering offers a novel avenue for introducing new traits for anthracnose tolerance. One of the mechanisms of conferring resistance to fungal pathogens is to introduce into the genome of plants genes that encode proteins such as chitinase and chitosanase that target components of fungal cell walls. Chitinase and chitosanase hydrolyse chitin and chitosan of fungal cell wall, render fungal cells osmotically sensitive and ultimately destroy the fungus (Cabib, 1987; Garcia et al., 1994). Chitinases endolytically hydrolyse the β-1,4-linkages of chitin while chitosanases on the other hand, hydrolyze the β-1,4-linkages between N-acetyl-D-glucosamine and D-glucosamine residues in a partially acetylated chitin polymer (Garcia et al., 1994; Lorito et al., 1998: Hendrix and Stewart, 2002). Glucosamine oligomers, released from fungal cell walls after hydrolysis with chitinase or chitosanase, are elicitors of plant defence response such as stomata closure (Lee et al., 1999) and cell wall lignifications (Vander et al., 1998). Mycoparasitic fungi such as *Trichoderma harzianum* are natural sources of fungitoxic proteins and produce chitinases and chitosanases among other hydrolytic proteins (Garcia et al., 1994; Lorito et al., 1998; Hendrix and Stewart, 2002). Co-expression of the *chitinase* (*harchit*) and *chitosanase* (*harcho*) genes from *T. harzianum* can therefore result in a synergistic enhancement of anti-fungal activity. The aim of this study was to transform sorghum with *harchit* and *harcho* genes encoding chitinase and chitosanase enzymes, respectively, to develop transgenic sorghum lines that are tolerant to fungal diseases. Sorghum immature zygotic embryos (IZE) were transformed by particle bombardment with the *harchit* and *harcho* genes and plantlets regenerated through tissue culture. The herbicide resistance gene encoding the enzyme phosphinothricin acetyltransferase (*pat)* was used as the selectable marker. *In planta* and *ex planta* infection assays were carried out to compare the response of the transgenic and wild type sorghum lines to *C. sublineolum* infection.

MATERIALS AND METHODS

Plant growth and pathogen culture

Sorghum lines were cultivated in the greenhouse under standard conditions consisting of a 16 hours/8 hours day/night photoperiod of 220 $\mp$ 20 $\mu E/m^2/s$ at changing day/night temperature of 24°C/20°C (Oldach et al.,

2001). Seedlings for *ex planta* and *in planta* infection assays were raised in controlled conditions (28°C, 220 μE/m²/s, 16hours/8 hours light/darkness) in a growth chamber. Stock cultures of *C. sublineolum* were maintained on 1/2 potato dextrose agar (PDA) (2.0 g/l potato starch, 10 g/l dextrose, 7.5 g/l agar). Sporulation was induced by growing 1/2 PDA-derived colonies on 7.25% (w/v) oatmeal agar (OMA) (Boora *et al.* 1998) at 26°C with 50 μE/m²/s backlight illumination for at least two weeks.

Vectors

The *chitinase (harchit)* (NCBI GeneBank Accession AB041751) and the *chitosanase (harcho)* (NCBI GeneBank Accession (AY571342.1) genes were extracted from *T. harzianum* and cloned into the pUbiCas plasmid (Dr. D. Becker, University of Hamburg, Germany, unpublished) and used to transform sorghum. 5.766 kbp (pUbiHarchit) plasmid vector was constructed by cloning *Harchit* into the pUbiCas plasmid under control of the constitutive *ubiquitin1* promoter (*ubi*1 promoter) from maize (*Zea mays*) (Christensen et al., 1992) and the nopaline synthase terminator (T*nos*) from *Agrobacterium tumefaciens* (Figure 1). The chitosanase gene *harcho* was cloned as described for *harchit* to form the vector pUbiHarcho (Figure 2). The p35SAcS vector containing *pat* gene encoding the enzyme phosphinothricin acetyltransferase (Strauch, 1988) for herbicide resistance was used as selectable marker (Figure 3).

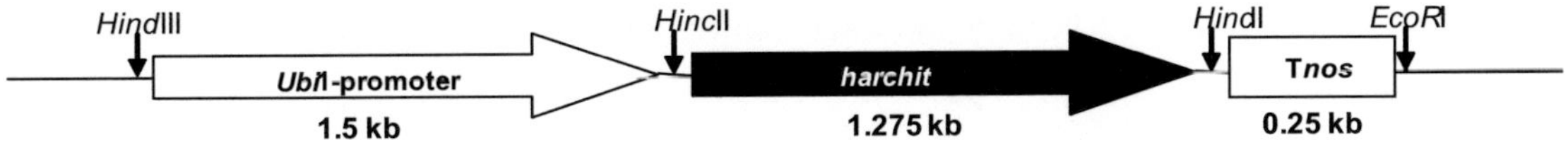

Figure 1. pUbi*HarChit* vector. The transformation vector contains *ubi*1 promoter, *harchit* and T*nos* terminator gene regions and restriction sites used in cloning and Southern Blot analysis.

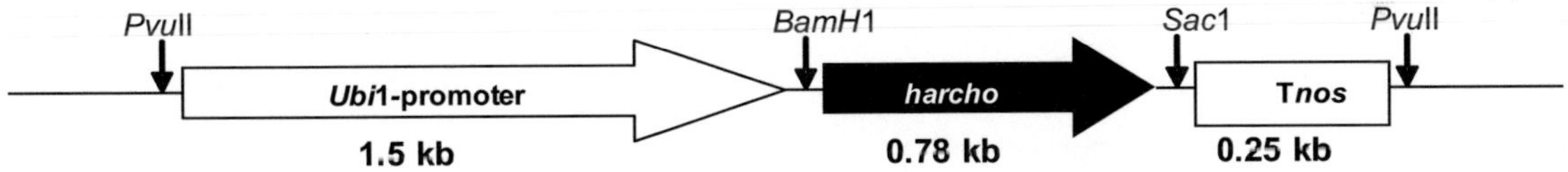

Figure 2. pUbi*HarCho* vector. The transformation vector contains *ubi*1 promoter, *harcho* and T*nos* terminator gene regions and restriction sites used in cloning and Southern Blot analysis.

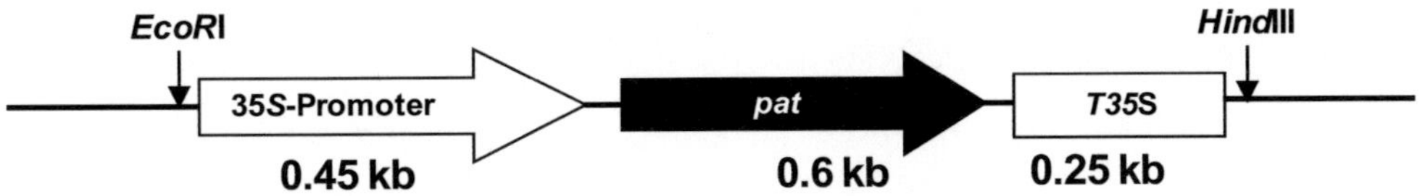

Figure 3. p35SAcS vector. The transformation vector contains CaMV 35S promoter, pat and T35S terminator genes and restriction sites used in cloning.

Isolation and genetic transformation of immature zygotic embryos

Immature seeds were harvested from sorghum panicles 14-20 days after flowering, washed 3 times with distilled water, sterilized with 2% sodium hypochlorite containing 0.1% MUCASOL® detergent (Brand Gmbh + Co KG, Wertheim, Germany) for 25 minutes, rinsed with sterile double distilled water. Immature zygotic embryos (IZE) were thereafter aseptically isolated from the immature sorghum seeds under a dissecting microscope. The IZE were transformed with gold particle coated with DNA plasmids containing *harchit*, *harcho*, *pat* genes through particle bombardment using the BIORAD PDS-1000/He Particle Gun (BioRad, München, Germany) (Brettschneider et al., 1997).

Tissue culture of IZE

The bombarded IZE were cultured in a MS-based (Murashige and Skoog 1962) callus induction (CIM), regeneration (REM) and rooting media (ROM), with requisite macroelements, microelements, vitamins, amino acids, recommended additives and plant hormones (Oldach et al., 2001). For induction of calli formation, sorghum IZE were cultured in darkness for two weeks at 26°C on CIM containing 2.5 mg/l 2,4-dichlorophenoxyacetic acid (2,4-D) and 0.1 mg/ml 6-benzylaminopurine (BAP). Calli were thereafter transferred onto a regeneration medium (REM) with 1 mg/l of BAP and cultured under 16/8 h light/darkness cycle at 26°C. The calli were cultured on REM for two weeks without selection and thereafter sub-cultured onto SEL-REM plates with 1 mg/l of BAP and 2 mg/l of the herbicide BASTA for 4-10 weeks. Calli and young regenerants were sub-cultured every 2-3 weeks onto fresh media. Regenerants with shoots were sub-cultured onto hormone-free ½ MS rooting-induction media with 2 mg/l BASTA. Regenerants were transferred to the greenhouse upon attaining a height of 7-10 cm.

Progeny segregation

Segregation of the transgenes in sorghum progeny was evaluated from the expression of the herbicide resistance gene. T_1 seedlings were sprayed twice with 200 mg/l and 300mg/l of the herbicide, BASTA at 7 and 14 days after germination, respectively. Surviving and dead plants were then enumerated and computed to establish the segregation ratio.

DNA isolation and Southern Blot analysis

Genomic DNA was extracted from sorghum leaves as described by Dellaporta *et al.* (1983). Digoxigenin-11-dUTP (DIG-dUTP) from Roche Diagnostics (Roche, Mannheim, Germany) was used to label DNA probes for detection of *harchit* and *harcho* genes in Southern blot analysis. DIG-labeling was done through PCR of respective plasmid DNA with *harchit* and *Harcho* DIG-primers (Table 1). The PCR reaction mix used for DIG-labeling included 0.3 μM of each Primers; 0.3 mM of each dATP, dGTP, dCTP; 0.2 mM dTTP; 35 μM DIG-dUTP; 2 mM $MgCl_2$, 1x PCR buffer; 1 U Taq-polymerase and 50 ng plasmid DNA. Southern blotting was carried out as described by Sambrook et al. (1989).

Table 1. Southern Blot Primers sequences for harchit and harcho showing position and PCR product length.

	Forward Primer	Primer position	Product Length
HarChitDig (5′-Primer):	5′-TCT CTG GCG ATA CCT ACG CT-3′	244-264	332bp
HarChitDig (3′-Primer)	5′-AGA TCG GAC TTC CTT CAG CA-3′	556-576	
HarChoDig (5′-Primer)	5′-TAG TGG CGG AGC TAC T-3′	135-151	381bp
HarChoDig (3′-Primer)	5′- GCT TGC TTT GGG TAA G-3′	500- 516	

RNA isolation, cDNA synthesis and qRT-PCR

Total RNA was extracted from sorghum leaves using the peqGOLD TriFast extraction protocol according to the manufacturer's instruction (PeqLab Biotechnologie, Erlangen, Germany). Extracted 10 µg total-RNA was treated with 10 U of RNA-free DNAse and RNAse inhibitor as described by reagents manufacturer (Fermentas Life Sciences, St. Leon Rot, Germany).

cDNA was synthesized using 18-mer oligonucleotide (Oligo(dT)$_{18}$) primer, dNTPs, RNAse inhibitor and the Moloney murine leukaemia virus reverse transcriptase (M-MuLV) as recommended by the reagents manufacturer (Fermentas Life Science, Leon Rot, Germany).

The sorghum actin gene (NCBI Accession no. X79378) was used as the internal standard to quantify the expression of *harchit* and *harcho*. 150-200 bp *harchit* and *harcho* primers were used in the qRT-PCR. A serial dilution of 1 ng-1 fg of cDNA were used as the templates in a qRT-PCR BIORAD ICYCLER® (Hercules, CA, USA) to determine the efficiencies of the *harchit* and *harcho* primers (Table 2). SYBR GREEN 1 with fluorescein (Eurogentec, Seraing, Belgium) was used for detection in qRT-PCR. The reaction mixtures and thermocycler program were developed from the protocol described by the manufacturer (Eurogentec, Seraing, Belgium).

Table 2. qRT-PCR Primers of target genes, their respective forward and reverse primers, optimal annealing temperatures, polymerization fragment length, primers efficiencies and NCBI accession numbers.

Gene	Forward Primer (5′ → 3′)	Reverse Primer (5′ → 3′)	Fragment Length (bp)	Primer Efficiency	NCBI Accession
Sorghum Actin	AGG CGC AGT CCA AGA GGG GTA	ATG GCG GGG GGT GTT GAA GGT	226	93%	X79378
T. harzianum Chitinase (*harchit*)	ACC CCA ATG CCA CCC CCTTCA A	TAC CGG CCT CCC AGC TTC CAC TT	174	99.37%	AB041751
T. harzianum Chitosanase *(harcho)*	GTG GCC AGA GCG AGA CT	TCA CCC CAG ATA CCA TAG AA	199	79.83%	AY571342

Ex planta and *In planta C. sublineolum* infection assays

Response of transgenic and Wt sorghum to pathogen infection was carried out with conidial suspension of *C. sublineolum* in water containing 0.01% Tween detergent. A conidia-count of $1x10^6$ conidia/ml was used in all infection experiments. The leaf segments were infected through point inoculation with 10µl of conidia and incubated at 26°C under 16/8 h light/dark conditions. *In planta* infection assay was done by spraying two-leaf stage seedlings with conidia suspension until run-off. Responses to infection were quantified at specific time interval between 0-144 HPI by quantifying the timing and severity of symptoms emergence and development. A sampling size of 30 seedlings was used for both *ex planta* and *in planta* experiments. The number of leaves showing symptoms was enumerated at specific intervals. The infection experiments were done in triplicates. Non infected controls were included in all experiments to monitor secondary infection.

RESULTS

Sorghum tissues culture

A collection of Kenyan sorghum genotypes were used in tissue culture (Figure 4A and 4B). Successful tissue culture from IZE of Kenyan cultivars, KAT 412, KAT 369X, KAT 487, SDSH 513, GBK 046820, Serena, KAT L5 (Line 5), ICSV III, GBK 046812, GBK 046842 and GBK 046844 was achieved. Genotype dependent variation in the responses of IZE in tissue culture was noted. Differences in the amount of phenolic compounds secreted into the tissue culture media were observed. Cultivars, such as KAT 412, KAT 487 were associated with low levels of secretion and tissue culture pigmentation (Figure 4C) whereas KAT L5, Serena, GBK 046844 secreted copious amount of phenolic compounds into culture media (Figure 4D). It was also noted that sorghum lines that secreted copious amount of pigments produced significantly less calli and regenerants than those associated with low level of pigmentation during tissue culture.

Figure 4. Sorghum kernel, seeds colour and phenolics secretion in regeneration media: A: Assortment of different sorghum cultivars: 1-GBK 046812, 2-GBK 046820, 3-KOSA-1 T_1, 4-KAT 412 Wt, 5-KAT 369X, 6-Gadam, 7-GBK 046844 and 8-KAT 487; B: seeds of different colours of some of the cultivars studied; C and D: tissue culture condition in 3 weeks old REM culture of two sorghum line secreting low (C) KAT 487 and high (D) KAT 046844 levels of pigments into the culture medium.

Analysis of transgenes integration, inheritance and expression in sorghum

Eight different sorghum genotypes were tested for transformation: Aralba, KAT 412, KAT 487, KAT L5, GBK 046820, ICSV III, SDSH 513 and Serena. From these, one genotype was successfully transformed with *harchit, harcho* and the selectable marker *pat* gene. Three BASTA herbicide resistant sorghum plants,

KOSA-1, KOSA-2 and KOSA-3, were successfully regenerated after microprojectile bombardment of IZE of KAT 412 (Figure 5). Stable integration of *harchit* and *harcho* transgenes was confirmed by Southern blot and qRT-PCR. The transgenic plants showed normal growth. Three generations of KOSA-1 and one generation of KOSA-2 and KOSA-3 were analysed. KOSA-1 produced normal and fertile T_1, T_2 and T_3 progenies. KOSA-2 and KOSA-3 produced normal T1 progenies.

Figure 5. Herbicide resistance test of sorghum. Test shows 3 plants before BASTA spraying (A) and 7 days after spraying (B). The transgenic plant (center, B) was resistant to the herbicide and was later confirmed to contain the transgenes *harcho* and *harchit*.

Progeny segregation ratio was determined from BASTA resistance of T1 seedlings of KOSA-1. Analysis of BASTA resistant T_1 progeny of KOSA-1, KOSA-2 and KOSA-3 showed that an average of 73.92% were transgenic, which corresponded to a Mendelian segregation ratio of a dominant gene. Southern blot analysis was used to confirm integration. Molecular analysis of the three primary transformants, KOSA-1, KOSA-2 and KOSA-3, showed that both *harchit* and *harcho* were integrated into the genome of the transgenic plants as exemplified by the *harchit* Southern blot below (Figure 6). Real time expression of *harchit and harcho* in KOSA-1 was quantified by qRT-PCR. Both genes were expressed in all the transgenic seedlings of KOSA-1 analyzed.

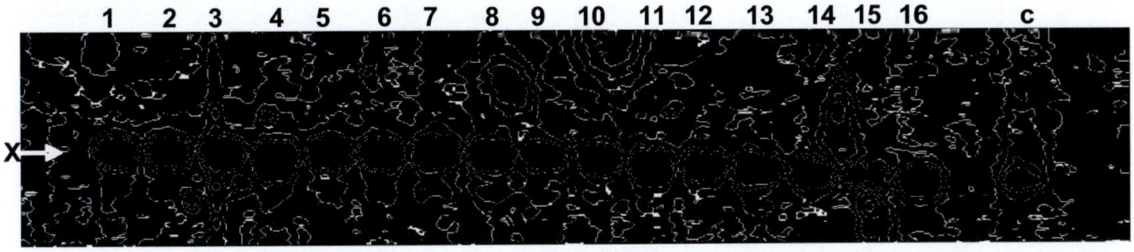

Figure 6. Southern blot of 3 generations of KOSA-1: Blot analysis for gene integration and inheritance of *harchit* gene in KOSA-1 T_o, (1), seedlings of T_1, (2-6), T_2 (7-11) and T_3 (12-16) and the positive control pUbiHarChit (c). The DNA of the KOSA-1 T_0-T_3 was restricted with *Hind*III and *Eco*RI enzymes that excised the promoter, *harchit* and *nos* gene cassette. The blots show inheritance of the *harchit* in successive generations.

Response to *C. sublineolum* infection

Response to infection was studied in the T_1 of the transgenic line, KOSA-1, parent Wt KAT 412 and 6 other Wt sorghum cultivars from Kenya: KAT 487, KAT L5, GBK 042820, GBK 042812, SGSH 513 and

Serena. *Ex Planta* and *In Planta* tests were done. Differences were noted in the timing and severity of symptoms and their development among the cultivars studied. Comparison of the transgenic KAT 412 T_1 (KOSA-1 T_1) and Wt KAT 412 showed that the transgenic plants developed less and late symptoms upon infection than the parent Wt plants (Figure 7). Comparison of transgenic KAT 412 T_1 with other cultivars established that there were differences among cultivars in the severity of infection symptoms (Figure 7). KAT L5 and SDSH 513 showed the least and most NLL at 48 HPI, respectively. By 72 HPI, most of the susceptible genotypes were symptomatic with KAT L5 having the least number of symptoms. Infection symptoms developed later in the transgenic KAT 412 T_1 than Wt, SDSH 513, Serena, and GBK 046820 but earlier than KAT L5, KAT 487 and KAT 046812. All the leaf segments from all the sorghum lines were symptomatic by day 5 (120 HPI). These observations established that KAT L5 and SDSH 513 were the most and least tolerant lines, respectively. The transgenic KAT 412 T_1 was more tolerant than the parent wild type, KAT 412 Wt. However, the transgenic line was less tolerant than the sorghum line KAT L5.

In planta infection assay was done on two weeks old seedlings. Seedlings were infected with spores of *C. sublineolum* and symptoms monitored over a period of 144 HPI. *In planta* assays revealed that the leaves of the parent KAT 412 Wt developed more and severe lesions than those of the transgenic KAT 412 T_1 (KOSA-1 T_1) seedlings (Figure 8).

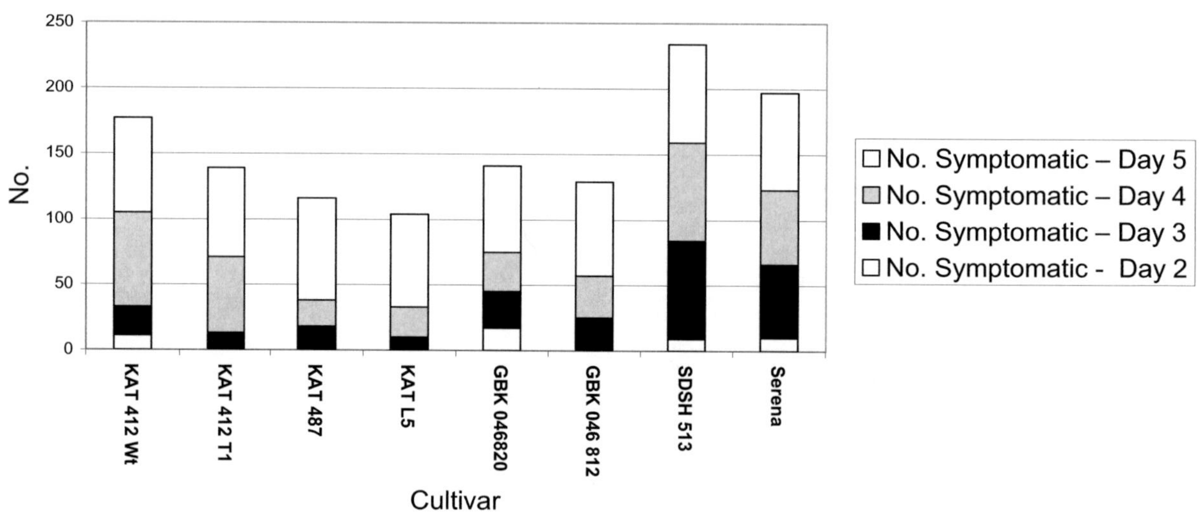

Figure 7. Comparison of *ex planta* response of Kenyan sorghum cultivars: Leaf segments were infected with 10µl of *C. sublineolum* and the number of segments showing symptoms was enumerated for 5 days. Transgenic plant (KAT 412 T1) was less symptomatic that the wild type KAT 412 Wt. Leaf segments of KAT L5 and SDSH 513 showed least and highest number of infection after 5 days post infection, respectively.

Comparison of the symptoms development 5 days after infection indicated that the leaves of the transgenic line were healthier than those of KAT 412 Wt (Figure 8). Response of the transgenic seedlings to *C. sublineolum* was further evaluated in comparison to sorghum cultivars, SDSH 513 and KAT L5. Disease symptoms in the seedlings of transgenic KOSA-1 were less severe in comparison to those of cultivar SDSH 513. However, seedlings of KAT L5 developed the least symptoms and showed the most tolerance as compared to SDSH 513, KAT 412 and KOSA-1 (Figure 9).

Figure 8. *In planta* sorghum infection assay: Seedlings were spray-infected with *C. sublineolum conidia*. Symptoms development was recorded on the 5th day (120 HPI) after infection. Virtually no symptoms was observed on KAT L5 leaves – resistant (A), few necrotic local lesions ((NLL) on KAT 412 T_1 leaves - tolerant (B), extensive NLL in KAT 412 Wt – less tolerant (C) and extensive NLL on both leaf blades and sheath, chlorosis and discolouration of both leaves of SDSH 513 – susceptible (D).

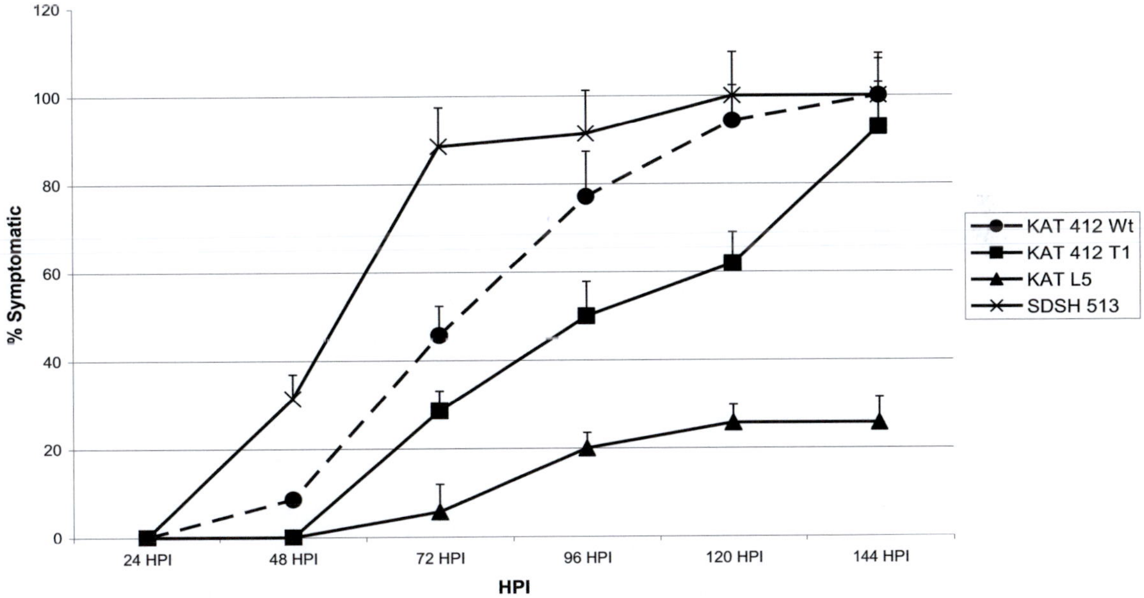

Figure 9. Symptoms development after *in planta C. sublineolum* infection: Sorghum seedlings were infected with conidia of *C. sublineolum* and the proportion of those showing symptoms enumerated between 24-144 HPI. Fewer KAT L5 (25%) developed symptoms later (72 HPI) than SDSH 513 that had more (100%) and sooner (48 HPI) infection. KOSA-1 was symptomatic later (72 HPI) and more tolerant than KAT 412 Wt. The transgenic line was more tolerant than the wild type parent.

The symptoms were quantified by computing the percentage of leaves showing symptoms every 24 hours for 144 HPI. There was variation among cultivar in the percentage of the leaves showing symptoms.

Quantification of the symptoms showed no symptoms were observable at 24 HPI, but an average 8.57% KAT 412 and 31.43% SDSH 513 were symptomatic at 48 HPI. Quantification of the symptomatic seedlings at 72 HPI showed that 5.71% KAT L5, 28.57% KOSA-1 T_1, 45.71% KAT 412 and 88.57 DSH 513 were symptomatic. It was notable that the transgenic line showed more tolerance than the parent Wt KAT 412. The percentages were subjected to square root transformation (√0-5+X) and statistically analysed with Newman-Keuls Test. ANOVA analyses showed that cultivar and time were statistically significant. The interaction of cultivar and time were also determined to be statistically significant (Table 4).

Table 4. ANOVA of square root transformation (√0-5+X) of the percentage of symptomatic leaves to establish the effect of cultivar and time on anthracnose symptoms development on four sorghum lines.

	Sum of Squares	Degrees of Freedom	Mean Square	F	Significance
Constant	6732.182	1	6732.182	426099.9	0.00*
Cultivar	603.041	3	201.014	12722.8	0.00*
Time	1920.701	5	384.140	24313.4	0.00*
Cultivar*Time	212.329	15	14.155	895-9	0.00*

* - significant at 0.05 confidence limit

This study established that the transgenic line KOSA-1 was more tolerant to anthracnose than the wild type KAT 412 parent line. It was also observed that the wild type genotype KAT L5 was the most tolerant while SDSH 513 was the least tolerant.

DISCUSSION

Tissue culture

Prolific tissue culture system is fundamental for successful regeneration of transgenic plants. A collection of sorghum genotypes sampled from Kenya were used to identify lines that could be consistently regenerated through tissue culture. A simple tissue culture protocol that comprised of 2 weeks calli induction, 2-4 weeks regeneration and 3-4 weeks rooting was used. Regeneration through somatic embryogenesis was achieved in all the cultivars studied. However, a genotype-dependent effect on calli development and plantlets regeneration ability was observed. The cultivars that secreted high levels of phenolics developed less calli and regenerants. Such genotype influence on somatic embryogenesis is observed (Casas et al., 1993; Takashi, 2002). It is also observed that sorghum explants, especially from cultivars that are rich in phenolic compounds are recalcitrant to morphogenesis during tissues culture (Cai and Butler 1990; Casas et al., 1993). Attempts have been made to ameliorate the negative effect of phenolics secretion into culture media. Anti-oxidants and anti-phenolics compounds and formulations such as polyvinylpyrrolidone (PVP), polyvinylpolypyrrolidone (PVPP), dithiothreitol, coconut water, silver nitrate and active carbon are used in sorghum tissue culture with mixed results (Zhao et al., 2000; Takashi, 2002). Tissue regeneration was achieved in sorghum lines both containing both low and high level phenolic compounds. This was achieved by limiting injuries to isolated embryos and maintaining the hormonal balance at 2.5 mg/l 2,4D and 0.1mg/l BAP in callus induction medium and 1 mg/l BAP in regeneration medium. The significance of the successful tissues culture of both phenolics-rich and poor lines lies in the potential use of the tissue culture protocol to tap into the genetic resource present in recalcitrant sorghum land races. This is important considering the fact that most of the recalcitrant high tannins and phenolics-rich brown-red sorghum lines are relatively tolerant to biotic and abiotic stress (Jambunathan, 1992; Esele et al., 1993; Waniska et al., 2001).

Genetic transformation

KAT 412 was successfully transformed with *harchit, harcho* and *bar* genes. Three fertile, herbicide resistant sorghum plants (KOSA-1, KOSA-2 and KOSA-3) were successfully developed through particle bombardment of IZE of KAT 412. Three generations of KOSA-1 and one generation of KOSA-2 and KOSA-3 showed normal growth and Mendelian segregation for herbicide resistance. Co-integration and expression of the two transgenes were confirmed by Southern blot and qRT-PCR analysis. Among the cereals, sorghum is one of the most difficult to transform (O´Kennedy et al., 2006). This is one of the first reports of successful co-transformation of sorghum through microparticle bombardment with two important anti-fungal genes: *harchit* and *harcho*. This a step forward in engineering sorghum for resistance against fungal diseases.

The production of transgenic sorghum plants via particle bombardment of IZE was reported for the first time by Casas et al. (1993) and subsequently by Zhu et al. (1998), Tadesse (2000, 2003), Able et al. (2001), Emani et al. (2002) and Girijashankar et al. (2005). Most of the early reports involved transformation with only reporter and marker genes such as *gus* and marker *bar* genes. Transformation of sorghum with an important gene was achieved by introducing a rice *chitinase* (Zhu et al., 1998; Krishnaveni et al. 2001) and a *Bacillus thuringiensis cry1Ac* coding a δ-endotoxin (Girijashankar et al., 2005). Transformation frequency in microparticle bombardment is traditionally low. Successful transformation of sorghum has been achieved with low frequencies: 0.008-0.33% (Casas et al., 1993, 1997) and 1.5% (Girijashankar et al., 2005). In this investigation three fertile transformants were developed. The low sorghum transformation frequency can be attributed to various factors that affect the response of sorghum to morphogenesis as well as innate characteristics that make the integration of foreign genes and regeneration of transformants difficult. The large amount of phenolic compounds produced by sorghum cultures is associated with the low transformation frequency (Cai and Butler, 1990; Casas et al., 1993). Transformation is affected by the physiological condition of the explant (Casas et al., 1993; Emani et al., 2002). Biotic and abiotic stress affects the transformation success (Casas et al., 1993; Emani et al., 2002). Artificial growth conditions in greenhouse marked by low illumination could have also affected the state and quality of the explant used for transformation.

Response to fungal infection

Ex planta and *in planta* assays were used to assess the response to *C. sublineolum* infection and to compare the tolerance level of the transgenic and wild type sorghum lines. In *ex planta* assay, leaf segments of sorghum seedlings were used. Similar methods are used for *in vitro* infection assays by Girgi et al. (2006) in pearl millet and Oldach et al. (2001) in wheat. Leaf segments from the transgenic KOSA-1 were found to be more tolerant to *C. sublineolum* and exhibited delayed and lower levels of necrosis after infection than those excised from the Wt parent plants. Comparison was also made between the transgenic and selected sorghum lines from Kenya. A statistically significant genotypic difference was noted on the timing and severity of the onset and development of disease symptoms. Sorghum cultivars KAT L5 and SDSH 513 were found to be the most tolerant and susceptible of the lines tested.

Effect of the transgenes on response to fungal infection

Infection assay of the transgenic and parent Wt sorghum lines showed that introduction of *harchit* and *harcho* into sorghum caused a significant level of protection against *C. sublineolum* under the conditions of this investigation. This study demonstrated the effectiveness of fungal cell wall-active proteins in protecting plants against these pathogens. The anti-fungal activity of *harchit* and *harcho* is confirmed in other crops (Lorito et al., 1998; Hendrix and Stewart, 2002). Selected transgenic tobacco and tomato lines containing antifungal endochitinases from the mycoparasitic fungus *T. harzianum* are highly tolerant to completely resistant to the foliar pathogens *Alternaria alternate*, *A. solani*, *Botrytis cinerea* and the *Rhizoctonia solani* (Lorito et al., 1998). The activity of *harcho* in activating defence mechanism is demonstrated in transgenic tobacco (Hendrix and Stewart, 2002). Chitinases from other sources are also known to be active against fungal pathogens. Zhu et al. (1998) and Krishnaveni et al. (2001) showed the protection of sorghum through the introduction of a rice *chitinase* gene. Barley class II *chitinase* also show activity against fungal pathogens (Punja, 2001). In addition to the chitinases and chitosanase, other antifungal proteins also enhance fungal disease tolerance and resistance in cereals and other crops. Transgenic wheat expressing resveratrol synthase genes *vst1* and *vst2* show a substantial reduction of disease symptoms caused by *Puccinia recondita f. sp. tritici* (Serazetdinova et al., 2005). Other cereals such as pearl millet (*Pennisetum glaucum*) have also been transformed with various antifungal proteins for fungal diseases resistance, for example antifungal protein (*afp)* from Aspergillus giganteus (Girgi et al., 2002; Girgi et al., 2006) and synthetic prawn protein encoding gene (*pin*) (Latha et al., 2006). In wheat, introduction of genes encoding antifungal protein Ag-AFP from the fungus *Aspergillus giganteus*, a barley class II chitinase inhibit the spread of *Puccinia recondita f. sp. tritici* and *Erysiphe graminis f. sp. tritici* (Oldach et al., 2001). This investigation contributed to the work on the development of disease tolerant cereals.

Conclusion

This work, taken into consideration with other sorghum and cereals transformations, demonstrates the potential of enhancing disease tolerance by combining (stacking) genes encoding antifungal proteins such as chitinase and chitosanase. The study also demonstrated the existence of genetic diversity in sorghum grown in Kenya that could be exploited for further transformation for anthracnose resistance.

ACKNOWLEDGEMENTS

This work was undertaken at Hamburg University, Department of Applied Plant Molecular Biology (AMPII) through a DAAD scholarship and a study leave from KIRDI. Gratitude to Prof. Dr. Horst Lörz for the research opportunity in AMPII, Louis Prom, PhD of USDA-ARS, College Station, TX, who provided the *C. sublineolum*, Calleb Olweny of Kenya Sugar Research Foundation (KESREF), Evans Mutegi of Kenya Agricultural Research Institute (KARI)-Genebank at Muguga, P.K. Kamau, Silas Nzioki and Rachel Maswili of KARI-Katumani who provided sorghum seeds.

REFERENCES

Able JA, Rathus C, Godwin ID. (2001). The investigation of optimal bombardment parameters for transient and stable transgene expression in sorghum. *In vitro* Cellular Development Biology Plant, 37:341–348.

Becker D, Brettschneider R, Lörz H. (1994). Fertile transgenic wheat from microprojectile bombardment of scutellar tissue. Plant J., 5:299-307.

Boora, KS, Frederiksen R, Magill C. (1998). DNA-based markers for a recessive gene conferring anthracnose resistance in sorghum. Crop Science, 38:1708-1709.

Brettschneider R, Becker D, Lörz H. (1997). Efficient transformation of scutellar tissue of immature maize embryos. Theor Appl Genet, 94:737-748.

Bueso, FJ, Waniska, RD, Rooney WL, Bejosano F. (2000). Activity of antifungal proteins against mold in sorghum caryopses in the field. Journal of Agricultural and Food Chemistry, 4:810-816.

Cabib E. (1987). The synthesis and degradation of chitin. In: Meisler A (ed) Advances in enzymology and related areas of molecular biology. Wiley, New York, pp 59-101.

Cai T, Butler L. (1990). Plant regeneration from embryogenic callus initiated from immature inflorescences of several high-tannin sorghums. Plant Cell Tissue Organ Culture, 20:101–110.

Casas AM, Kononowicz AK, Zehr UB, Tomes DT, Axtell JD, Butler LG, Bressan RA, Hasegawa PM. (1993). Transgenic sorghum plants via microprojectile bombardment. Proceedings of the National Academy of Sciences USA, 90:11212–11216.

Casas AM, Kononowicz AK, Haan TG, Zhang L, Tomes DT, Bressan RA, Hasegawa PM. (1997). Transgenic sorghum plants obtained after microprojectile bombardment of immature inflorescences. *In vitro* Cell Development Biology-Plant, 33:92–100.

Christensen AH, Sharrock RA and Quail PH. (1992). Maize polyubiquitin genes: structure, thermal perturbation of expression and transcript splicing, and promoter activity following transfer to protoplasts by electroporation, Plant Molecular Biolology, 18:675–689.

Dellaporta SL, Wood J, Hicks JB. (1983). A plant DNA minipreparation: version II. Plant Molecular Biolology Rep 1:19–21.

Emani C, Sunilkumar G, Rathore KS. (2002). Transgene silencing and reactivation in sorghum. Plant Science 162:181–192.

Esele JP, Frederiksen RA, Miller FR. (1993). The association of genes controlling caryopsis traits with grain mold resistance in sorghum. Phytopathology 83:490-495

García I, Lora JM, De la Cruz J, Benítez T, Lobell A, Pintor-Toro JA. (1994). Cloning and characterization of a chitinase (CHIT42) cDNA from the mycoparasitic fungus *Trichoderma harzianum.* Current Genetics, 27:83-89.

Gebeyehu G, Adugna A, Tadesse T, Tesso T, Belete K, Michael H. (2004). Development of sorghum varieties and hybrids for dryland areas of Ethiopia. Uganda Journal of Agricultural Science, 9:594-605.

Girgi M, O'Kennedy MM, Morgenstern A, Smith G, Lörz H, Oldach KH. (2002). Transgenic and herbicide resistant pearl millet (*Pennisetum glaucum* L.) R.Br. via microprojectile bombardment of scutellar tissue. Molecular Breeding, 10:243–252.

Girgi M, Breese WA, Lörz H, Oldach KH. (2006). Rust and downy mildew resistance in pearl millet (Pennisetum glaucum) mediated by heterologous expression of the afp gene from Aspergillus giganteus. Transgenic Research, 15:313–324.

Girijashankar V, Sharma HC, Sharma KK, Swathisree V, Prasad LS, Bhat BV, Royer M, Secundo BS, Narasu ML, Altosaar I, Seetharama N. (2005). Development of transgenic sorghum for insect resistance against the spotted stem borer (*Chilo partellus*). Plant Cell Reports 24:513–522.

Hammer JE, Howard RJ, Chumley FG, Valent B. (1988). A mechanism for surface attachment in spores of a plant pathogenic fungus. Science, 239:288-290.

Hendrix B, Stewart J Mc D. (2002). Chitosanse may enhance anti-fungal defence response. AAES Research Series 507:2154-217.

International Crops Research Institute for the Semi-Arid Tropics (ICRISAT). (2010). An Endowment Fund for Sorghum and Pearl Millet Research Enhanced nutritional security, biodiversity and environmental sustainability. www.icrisat.org/Investors/Flyers/59Endowment%20fund_D_scr.pdf.

Jambunathan R, Kherdekar MS, Stenhouse JW. (1992). Sorghum grain hardness and its relationship to mold susceptibility and mold resistance. Journal of Agricultural and Food Chemistry, 40:1403–1408.

Krishnaveni S, Jeoung JM, Muthukrishnan S, Liang GH. (2001). Transgenic sorghum plants constitutively expressing a rice chitinase gene show improved resistance to stalk rot. Journal of Genetics and Breeding, 55:151–158.

Latha, AM, Rao KV, Reddy TP, Reddy VD. (2006). Development of transgenic pearl millet (*Pennisetum glaucum* (L.) R. Br.) plants resistant to downy mildew. Plant Cell Reports, 29:927–935.

Lee S, Choi H, Suh S, Doo I, Oh K, Choi EJ, Taylor ATS, Low PS and Lee Y. (1999). Oligogalacturonic acid and chitosan reduce stomatal aperture by inducing the evolution of reactive oxygen species from guard cells of tomato and *Commelina communis*. Plant Physiology, 121:147-152.

Lorito M, Woo SL, Garcia I, Colucci G, Harman GE, Pintor-Toro JA, Filippone E, Muccifora S, Lawrence CB, Zoina A, Tuzun S, Scala F. (1998). Genes from mycoparasitic fungi as a source for improving plant resistance to fungal pathogens. Proceedings of the National Acadaemy of Sciences, USA, 95(14):7860-7865.

Machado S, Bynum ED, Archer TL, Bordovsky J, Rosenow DT, Peterson C, Bronson K, Nesmith DM, Lascano RJ, Wilson LT, Segarra E. (2002). Spatial and Temporal Variability of Sorghum Grain Yield: Influence of Soil, Water, Pests, and Diseases Relationships. Precision Agriculture, 3(4):389-406(18)

Murashige T, Skoog F. (1962). A revised medium for rapid growth and bioassays with tobacco tissue cultures. Physiol Plant, 15(3):473-497.

Nicholson RL, Epstein L. (1991). Adhesion of fungi to the plant surface: Prerequisite for pathogenesis. In: Cole, G. T., and Hoch, H. C., eds., The fungal spore and disease initiation in plants and animals. Plenum Press, New York.

Oldach KH, Morgenstern A, Rother S, Girgi M, O'Kennedy MM, Lörz H. (2001). Efficient in vitro plant regeneration from immature zygotic embryos of pearl millet [*Pennisetum glaucum* (L.) R. Br.] and *Sorghum bicolor* (L.) Moench. Plant Cell Reports, 20:416–421.

Punja Z. (2001). Genetic engineering of plants to enhance resistance to fungal pathogens — a review of progress and future prospects. Canadian Journal of Plant Pathology, 23:216–235.

Prom LK, Waniska RD, Kollo AI, Rooney WL, Bejosano FP. (2005). Role of Chitinase and Sormatin Accumulation in the Resistance of Sorghum Cultivars to Grain Mold. Journal of Agricultural and Food Chemistry, 53(14):5565-5570.

Sambrook J, Fritsch EF, Maniatis T. (1989). Molecular cloning: a laboratory manual, 2nd edn, Cold Spring Habour Laboratory, Cold Spring Harbour, New York.

Serazetdinova L, Oldach KH, Lörz H. (2005). Expression of transgenic stilbene synthases in wheat causes the accumulation of unknown stilbene derivatives with antifungal activity. Journal of Plant Physiology, 162(9):985.

Strauch E, Wohlleben W, Pühler A. (1988). Cloning of a phosphinothricin N-acetyltransferase gene from *Streptomyces viridochromogenes* Tü494 and its expression in *Streptomyces lividans* and *Escherichia coli*. Gene 63:65-74

Sutton BC. (1980). The Coelomycetes: Fungi Imperfecti with Pycnidia, Acervuli and Stromata. Commonwealth Mycological Institute, Kew, London, 696 pp.

Tadesse YS. (2000). Genetic transformation of sorghum (*Sorghum bicolor* (L.) Moench) towards improving nutritional quality. Ph.D. thesis. Vrije Universiteit Brussel Instituut Voor Moleculaire Biologie, Faculteit Wetenschappen, Belgium.

Tadesse Y, Sgi L, Swennen R, Jacobs M (2003) Optimisation of transformation conditions and production of transgenic sorghum (*Sorghum bicolor*) via microparticle bombardment. Plant Cell, Tissue and Organ Culture, 75:1–18.

Takashi H. (2002). Adventitious shoot regeneration from immature embryos of sorghum. Plant Cell, Tissue and Organ Culture, 68(1):65-72.

Thomas MD, Sissoko I, Sacko M. (1996). Development of leaf anthracnose and its effect on yield and grain weight of *sorghum* in West Africa. Plant Disease, 89:151-153.

Vander P, Vrum KM, Domard A, Eddine El Gueddari N, Moerschbacher BM. (1998). Comparison of the ability of partially N-acetylated chitosans and chitooligosaccharides to elicit resistance reactions in wheat leaves. Plant Physiology, 118(4):1353-9.

Waniska RD, Venkatesha RT, Chandrashekar A, Krishnaveni S, Bejosano FP, Jeoung J, Javaraj J, Muthukrishnan S, Liang GH. (2001). Antifungal proteins and other mechanisms in the control of sorghum stalk rot and grain mold. Journal of Agricultural and Food Chemistry, 49(10):4732-42.

Wenzel WG, Rooyen PJ. (2001). Moisture stress and potential sorghum yield. International Sorghum and Millets Newsletter 42:28-29.

Zhao ZU, Cai T, Tagliani L, Miller M, Wang N, Pang H, Rudert M, Schroeder S, Hondred D, Seltzer J, Pierce D. (2000). *Agrobacterium*-mediated sorghum transformation. Plant Molecular Biology, 44:789–798.

Zhu H, Muthukrishnan S, Krishnaveni S, Wilde G, Jeoung JM, Liang GH. (1998). Ballistic transformation of sorghum using a rice chitinase gene. Journal of Genetics and Breeding, 52:243–252.

18. Development and evaluation of briquettes from selected agricultural by-products

Daniel Kerich[1*], Jackis Aukah[2] and Julius Ogola[3]
[1]Energy Division, KIRDI Western Region Centre (KWRC), Kisumu, Kenya.
[2]Engineering Division, KIRDI Western Region Centre (KWRC), Kisumu, Kenya.
[3]Laboratory Services Center, KIRDI Western Region Centre (KWRC), Kisumu, Kenya.
Corresponding author: dankerych@gmail.com

INTRODUCTION

The importance of energy in national development cannot be overemphasized as this can contribute immensely to economic and social life of a nation. The level and the intensity of energy use in a country is a key indicator of economic growth and development. The Kenya Vision 2030 (GoK, 2007) identified energy as one of the infrastructure enablers of its social economic pillar. Sustainable, affordable and reliable energy for all citizens is a key factor in realization of the Vision.

Wood fuel and its derivative (charcoal) supplies about 95% of the domestic and commercial cooking energy needs in developing countries. Since most developing countries' economies are primarily agriculturally based, they produce huge quantities of agricultural residues - such as wheat and rice straws, maize, millet and sorghum stalks, rice husks and cassava peel - which are a potential source of cheap energy. Agricultural residues are not attractive as a household fuel source because they are bulky and have low energy intensity. To eliminate smoke generation when burning, agricultural residues require processing by carbonization before being used as a house-hold indoor fuel. These residues can be converted to biomass energy by briquetting them prior to use. Briquetting is a process of compaction of residues into a product of higher density than the original material. According to Grover and Tabil (1997), briquetting process can be classified under two broad groups, which are briquettes without binder and briquettes with binding agent. In terms of technique of briquetting, there are low-pressure and high-pressure briquetting processes (Olorunnisola, 2007).

Briquetting of agricultural by-products can promote their economic utilization whilst being a solution to local energy shortages and replacing charcoal as a source of energy. The aim of this work was to develop, evaluate and investigate the fuel properties of briquettes produced from rice husks and other agricultural by-products, with clay paste, cassava starch or molasses as binder material. Rice husk is a typical dry or membranous outer covering of the rice. When compared to other agricultural by-products, it has higher ash and potash content (3.84%). Rice husk is an exceptional biomass because it has good flow ability; low moisture content (10-12%) and the ash contain fewer alkaline minerals. It makes an excellent fuel although its calorific value is less than wood and other agro residues.

MATERIALS AND METHODS

Sources and preparation of materials

Rice husk (7-8% moisture content) was obtained from Lake Basin Development Authority Rice Mill, Kisumu. Bagasse (8-10% moisture content) and liquid molasses were obtained from Kibos Sugar and Allied Company, Kisumu. Saw dust (8-10% moisture content) was obtained from Kibuye market saw mill operators, Kisumu. Cassava starch (9-10% moisture content) was obtained from Kibuye market, Kisumu. Clay was obtained from Kisumu town.

The raw biomass materials were first analyzed for proximate composition and then carbonized in a kiln. Carbonized materials together with binder were properly mixed in a trough to ensure even mixing. The mixture was fed into the briquetting machine/mould for compaction. The compacted material was sun-dried for 2-3 days. Briquettes were made with either 10% cassava starch, 15% clay or 20% molasses w/w as binder material. The briquettes were produced using an electrical screw type briquetting system (Figure 1) and a manual briquetting machine (Figure 2, Figure 3) developed in KIRDI.

Figure 1. Briquetting machine

Figure 2. Parts of a manual press

Figure 3. Manual press

Cassava porridge was prepared by adding 20g cassava starch in 250 ml water. This was cooked to obtain a gel then cooled. Molasses was prepared by diluting it with water at the ratio of 1:2. Clay paste binder was prepared by dissolving 35g of dried clay in 250ml water. Ash content, moisture content, fixed carbon and calorific value of both raw material samples and briquettes were analyzed in accordance with ISO standards for analysis of solid fuels (EN 1860-2:2005). The ISO standard for each procedure is given below for each of the tests carried out. The other briquette parameters that were measured were shatter test, bulk density, resistance to water penetration test, degree of densification test and energy density ratio.

ANALYSIS AND EVALUATION OF BRIQUETTES

Moisture content

Moisture content was determined in accordance with IS0 579 in which the moisture content of raw biomass (1-2 g) and briquetted fuel was determined by calculating the loss in weight of material using hot air oven at 105°C for 1 hour and up to constant weight loss. The dish was removed from the hot air oven and put in a desiccator to cool. After cooling the dish together with the sample was weighed. The moisture content M of the briquette charcoal as analysed, expressed as a percentage by mass, is given by the formula.

$$M = \frac{(m_1 - m_4) - (m_3 - m_2)}{(m_1 - m_4)} x100$$

Where

M is the moisture content, in percentage;

m_1 is the mass of the container plus sample in g;

m_2 is the mass of the dry, empty dish, in g;

m_3 is the mass of the dish plus sample after heating, in g;

m_4 is the mass of the dry, empty container, in g;

The mean result of triplicate determinations was reported to the nearest 0.1%. The dried material was used for volatile matter and ash content determination.

Volatile matter

The sample was placed in the crucible covered with a lid and placed in an electric furnace (muffle furnace), maintained at 925±20°C for 7 minutes (ISO 562). The crucible was cooled first in air, then inside a desiccator and weighed again. Loss in weight was reported as volatile matter on percentage basis. The volatile matter V on a dry basis expressed as a percentage by mass, is given by the equation

$$V = \frac{(m_2 - m_3)}{(m_2 - m_1)} x100$$

where

V is the volatile matter, as a percentage by mass, in the sample analysed;

m_1 is the mass of the empty crucible and lid, in g;

m_2 is the mass of the crucible and lid and sample after drying at 105° C for 1 hour, in g;

m_3 is the mass of the crucible and lid and content after heating, in g;

The mean of triplicate determinations was reported to the nearest 0.1%.

Ash content

By modifying ISO 1171 the residual sample in the crucible was heated without lid in a muffle furnace at 700±50°C for 30 minutes. The crucible was taken out, cooled first in air, then in desiccators and weighed. Heating, cooling and weighing was repeated till a constant weight was obtained. The residue was reported as ash on percentage basis. The percentage of ash was calculated from the mass of the residue after incineration. The ash *A* of the sample as analysed, expressed as a percentage by mass, on a dry basis is given by the formula

$$A = \frac{(m_3 - m_1)}{(m_2 - m_1)} x100$$

where

A is the ash content in percentage;

m_1 is the mass of the dish plus lid, in g;

m_2 is the mass of dish plus lid plus sample after drying at 105° C to 110°C for 1 hour, in g;

m_3 is the mass of dish plus lid plus ash, in g;

The mean of triplicate determinations was reported to the nearest 0.1%.

Fixed carbon

The fixed carbon percentage was calculated by using the following relationship:

Percentage of fixed carbon = 100 - % of (moisture content+ volatile matter + ash) wet basis.

Calorific value

The calorific value of the raw materials and the charcoal briquettes were determined in accordance with ASTM E 711-87 (2004). The apparatus used was a bomb calorimeter. The calorific value of biomass material was determined by using the following formula.

$$\textit{Calorific value (kcd/kg)} = \frac{(W + w)\ x\ (T_1 - T_2)}{x}$$

Where:

W = weight of water in calorimeter (kg)

w = water equivalent of apparatus

T_1 = initial temperature of water (°C)

T_2 = final temperature of water (°C)

X = weight of fuel sample taken (kg)

Bulk density

The bulk density of the briquettes was determined by measuring dimensions and weighing 3 samples. Weighing was performed by using an analytical balance and the dimension measured by using a vernier caliper. Three samples were tested for each experiment and the average density was reported.

Shatter indices

The drop test was conducted according to the ASTM D440 - Standard Method of Drop Shatter Test for Coal. Two samples from each test were used and dropped from a height of 2 m. Shatter indices were used to determine the hardness of briquettes. The briquette with known weight and length was dropped on reinforced cement concrete (RCC) floor. The weight of disintegrated briquette and its size was noted. The percent loss of material was calculated. The shatter resistance of the briquettes was calculated using following formula (Ghorpade and Moule, 2006).

$$Percent\ weight\ loss\ (\%) = \frac{(W_1 + W_2)}{W_2} x\, 100$$

% Shatter resistance = 100 - % weight loss

Where, w_1 = Weight of briquette before shattering

w_2 = Weight of briquette after shattering

Porosity index

Each briquette was immersed in a beaker containing a known volume of water for 20 minutes. The porosity of the briquette was determined based on the amount of water each was able to absorb. The porosity index was calculated as the ratio of the mass of water absorbed to the mass of the sample immersed in the water.

$$Porosity\ index,\ P_i = \frac{M_w}{M_b}$$

Where P_i – Porosity index

M_w – Mass of water absorbed

M_b – Mass of the sample briquette immersed in water

Water boiling test

This was carried out to compare the cooking efficiency of the briquettes. It measured the time taken for each set of briquettes to boil an equal volume of water under similar conditions. Each briquette sample (100 g) was used to boil 100 cm^3 of water using small steel cups and domestic briquettes stove. During this test, other fuel properties of the briquettes like burning rate and specific fuel consumption were determined. Also the level of smoke evolution was observed.

$$Burning\ rate,\ B_R\ (g/\min) = \frac{M_f}{T_{\min}}$$

Where M_f – Mass of the fuel burned in grams

T_{min} – Total time in minutes taken

$$Specific\ fuel\ consumption\ (sfc) = \frac{M_{fc}}{Q_w}$$

Where M_{fc} – Mass of fuel consumed in grams

Q_w – Quantity of boiling water in litres

Ignition time

Each briquette sample was ignited at the base with a cigarette lighter in a drought free place. The time required for the flame to ignite the briquette was recorded as the ignition time using a stop watch.

RESULTS AND DISCUSSION

Proximate analysis of biomass materials

The residues used in the study were first dried to ease carbonization hence the low moisture content ranging between 7-9%. Volatile matter was highest in bagasse at 80% and lowest in rice husk at 54% (Figure 5). Higher volatile matter is known to increase the ignition capability of a fuel material. Rice husk recorded the highest level of ash at about 25% (Figure 5). Bagasse had 3% ash while saw dust and coconut each had less than 1%. This makes rice husk unattractive as a fuel source because too much ash is problematic since it causes clogging in most firing appliances. Bagasse had 8% fixed carbon compared to saw dust and rice husk which both had about 15% (Figure 5). Fixed carbon was noted to be relatively lower with raw biomass compared to carbonized biomass residues. It can be concluded that the briquettes from bagasse and saw dust had better positive attributes than rice husk briquette, because the former had higher values of volatiles and lower percentage of ash content.

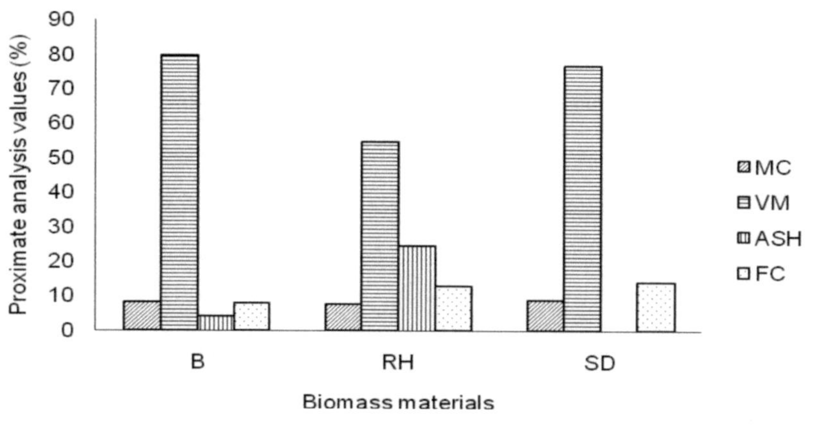

Figure 5. Proximate analysis of biomass. B - bagasse, RH - rice husk, SD - saw dust, MC - moisture content, FC - fixed carbon, VM - volatile matter and ASH- ash.

Calorific values of biomass materials

Rice husk had the lowest calorific value (figure 6, 3,084 kcal/g). This is mostly attributed to the high ash level of rice husk which substantially lowers the calorific value of the briquette. As shown in Figure 5 rice husk had 24.66% ash compared to bagasse with 4.12% and saw dust with 0.5%.

The net heating value calculated for rice husk was 3,084 kcal/kg while those of other residues were: bagasse 3,643 kcal/kg and sawdust 4,050 kcal/kg. These energy values are sufficient enough to produce heat required for household cooking and small scale industrial cottage applications. They also compare well with most biomass energy. For example, groundnut shell briquette – 3,009 kcal/kg (Musa, 2007), cowpea – 3,432 kcal/kg and soybeans – 3,093.77 kcal/kg (Enweremadu et al., 2004). The heating values of biomass residues from the other materials are higher than that of rice husk which shows that briquettes made from bagasse, sawdust and coconut residues were better than rice husk briquettes.

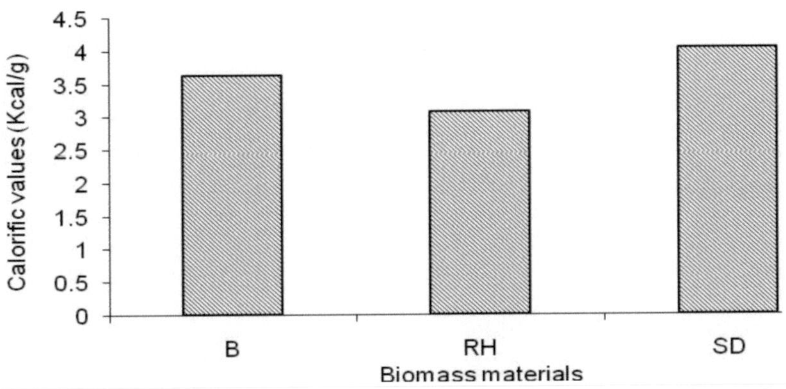

Figure 6. Calorific values of biomass material. B - bagasse, RH - rice husk, SD - saw dust.

Moisture content of briquettes

The moisture content of the briquettes was determined after 5 days drying in the sun. Rice husk recorded the least values of moisture content with all the binders. For the respective binders it was noted (Figure 7) that for clay paste binder, after five days rice husk reached 2.29% moisture content compared to bagasse which had 3.97% moisture content while saw dust had 4.03%. For other binders dried over the same days the results as indicated in Figure 7 were; cassava binder (rice husk 1.8%, bagasse 3.22%, sawdust 6.33%), molasses binder (rice husk 1.83%, bagasse 2.17%, sawdust 3.74%). Figure 7 indicate that rice husk gave low moisture content with all binders. This is a positive attribute for the rice husk briquettes for they can dry in a few days. An important observation to note is that molasses is hygroscopic and a few days after drying when left in the open the briquettes with molasses binder tend to absorb moisture and thus raise their moisture level. The moisture content analysis as captured in the figure was however done soon after drying the briquettes.

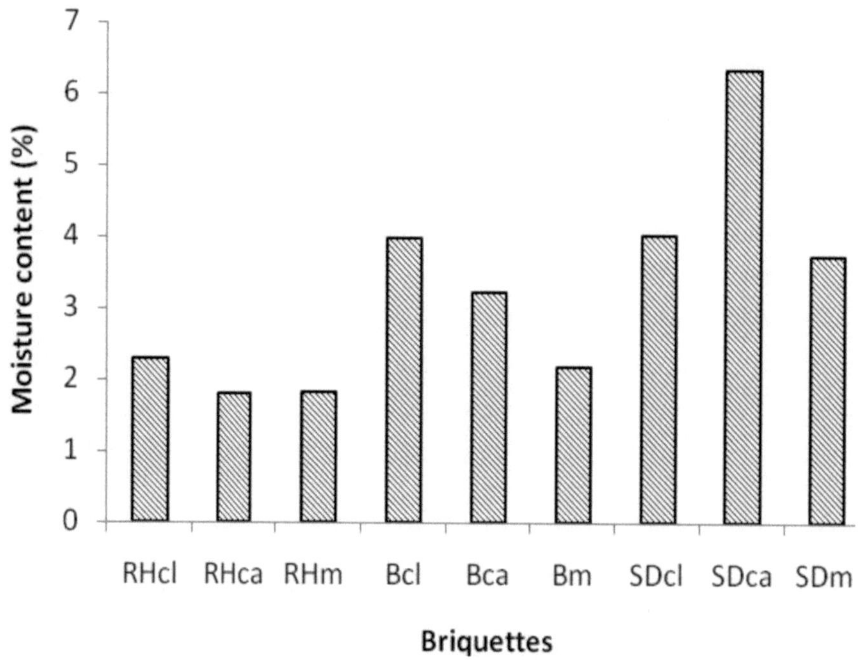

Figure 7. Moisture content (%) of briquettes. RH_{cl} – rice husk with clay binder, RH_{ca} – rice husk with cassava binder, RH_m – rice husk with molasses binder, B_{cl} – bagasse with clay binder, B_{ca} – bagasse with cassava binder, B_m – bagasse with molasses as a binder, SD_{cl} – saw dust with clay binder, SD_{ca} – saw dust with cassava binder, SD_m – saw dust with molasses.

Volatile matter of briquettes

Volatile matter is the gaseous component that comes out during combustion and is taken as the volatile solids fraction of the total solids. Less volatile solids are preferred in briquettes. Higher volatile solids reduce the calorific value and burning time. Carbonized residues had low volatile matter (< 47%, Figure 8) since most of it was driven out during carbonization and pyrolysis. Cassava and molasses binders contributed significantly to the reflected volatile matter in Figure 8. The two binders indicated higher values of volatile matter than clay in all the briquettes. For the three biomass materials, briquettes with molasses as the binder depicted the highest values followed by briquettes with cassava binder for the respective biomass materials. The elevated values are attributed to the binder materials since they are raw biomass and are rich in volatile matter. On the other hand, clay is mainly composed of inorganic material and has low volatile matter. Generally rice husk recorded lower values of volatile matter for all the three binders as compared to bagasse and saw dust.

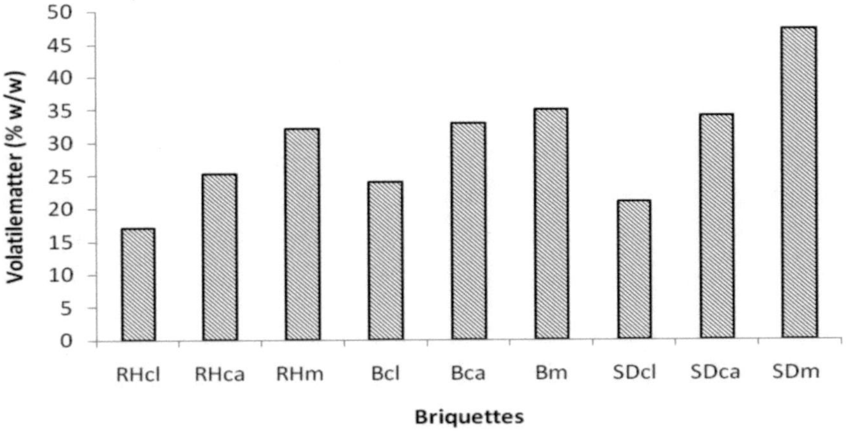

Figure 8. Volatile matter of briquettes. RH_{cl} – rice husk with clay binder, RH_{ca} – rice husk with cassava binder, RH_m – rice husk with molasses binder, B_{cl} – bagasse with clay binder, B_{ca} – bagasse with cassava binder, B_m – bagasse with molasses as a binder, SD_{cl} – saw dust with clay binder, SD_{ca} – saw dust with cassava binder, SD_m – saw dust with molasses.

Ash content of briquettes

The ash content of the briquette is the residue that remains when the carbonaceous portion is burned off. The ash consists of silica, aluminium, iron, magnesium and calcium. Ash is not desirable and is considered an impurity. Usually materials with low ash contents have high calorific values. Briquettes with clay binder recorded higher ash values (Figure 9) because clay is mostly incombustible. This generally lowered the calorific values of the briquettes but on the other hand increased the burning time of the briquettes. The other two binders, cassava and molasses are combustible and their ash content was low (<47%) in all briquettes compared to clay which had ash content >55%. For the three briquettes, with cassava and molasses binders, rice husks recorded slightly higher values of ash.

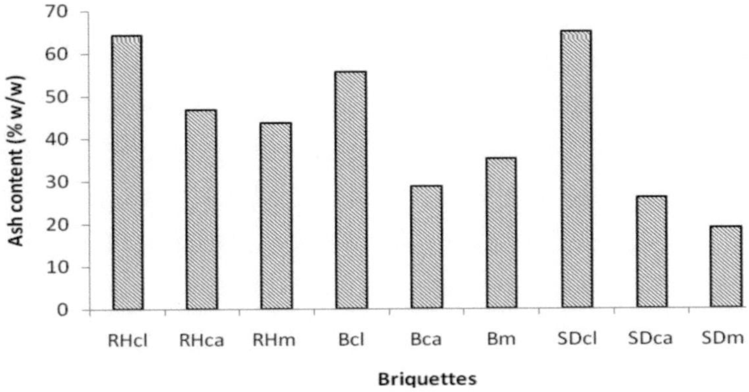

Figure 9. Ash content of briquettes. RH_{cl} – rice rusk with clay binder, RH_{ca} – rice husk with cassava binder, RH_m – rice husk with molasses binder, B_{cl} – bagasse with clay binder, B_{ca} – bagasse with cassava binder, B_m – bagasse with molasses as a binder, SD_{cl} – saw dust with clay binder, SD_{ca} – saw dust with cassava binder, SD_m – saw dust with molasses.

Fixed carbon content of briquettes

Fixed carbon represents the combustible material (carbon) in the briquette. From Figure 10, it was noted that briquettes from the three biomass materials yielded higher values with cassava as the binder material. The figures for fixed carbon with cassava for the three biomass materials were RHca (26.14%), Bca (35.3%) and SDca (33.65%). For molasses binder the values were also relatively high; RHm (22.62%), Bm (27.77%) and SDm (30.22%). Briquettes with clay as the binder material gave the lowest values for fixed carbon. From the analysis it was found that cassava used as the binder also improved the briquette quality in terms of fixed carbon.

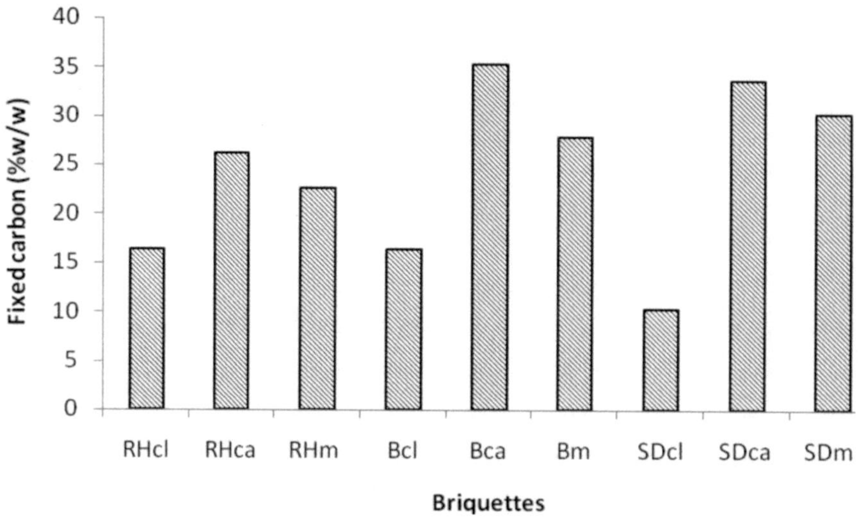

Figure 10. Fixed carbon of briquettes. RH_{cl} – rice husk with clay binder, RH_{ca} – rice husk with cassava binder, RH_m – rice husk with molasses binder, B_{cl} – bagasse with clay binder, B_{ca} – bagasse with cassava binder, B_m – bagasse with molasses as a binder, SD_{cl} – saw dust with clay binder, SD_{ca} – saw dust with cassava binder, SD_m – saw dust with molasses.

Calorific value of briquettes

The calorific value gives the energy obtained from the briquettes. The calorific values for the charcoal briquettes produced from the three biomass residues and different binders are shown in Figure 11. Briquettes with cassava and molasses binder indicated higher calorific values compared to briquettes with clay binders. For briquettes with cassava the values were; RHca (6,062 kcal/kg), Bca (7,150 kcal/kg) and SDca (7,056 Kcal/kg). These were also comparable to briquettes with molasses; RHm (6,024 kcal/kg), Bm (6,581 kcal/kg), SDm (7,275 kcal/kg). These values compared well with the calorific value of wood charcoal which is about 7,500 kcal/kg. Briquettes with clay binder generally showed lower calorific values; RHcl (4,896 kcal/kg), Bcl (5,183 kcal/kg) and SDcl (4,524 kcal/kg). Molasses and cassava are organic materials and therefore they combust well when used as binders and thus have no significant influence on the calorific values of the briquettes. If anything, they increase the calorific values. This is however not the case with clay as it is an inorganic material, which increases the ash percentage in the briquettes. High ash lowers the calorific value of the briquettes.

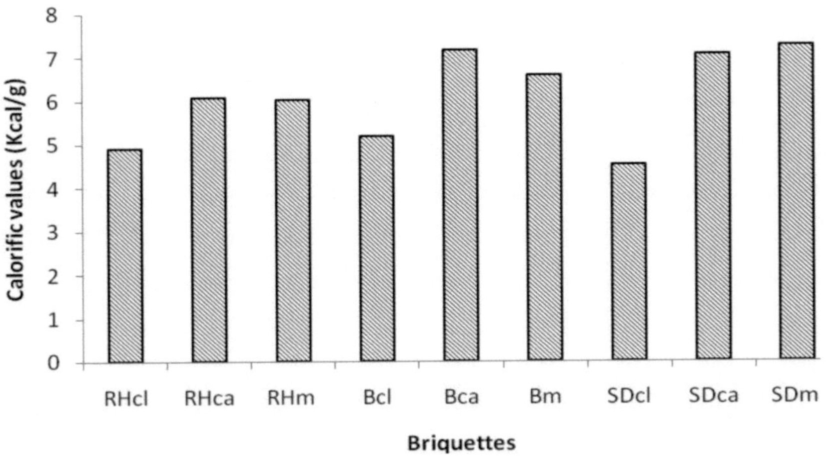

Figure 11. Calorific values of briquettes. RH_{cl} – rice husk with clay binder, RH_{ca} – rice husk with cassava binder, RH_m – rice husk with molasses binder, B_{cl} – bagasse with clay binder, B_{ca} – bagasse with cassava binder, B_m – bagasse with molasses as a binder, SD_{cl} – saw dust with clay binder, SD_{ca} – saw dust with cassava binder, SD_m – saw dust with molasses.

Bulk densities of briquettes

Density is particularly important in briquettes where high amount of material is combusted in a small volume. Bulk density is important when carbon is mixed with binders because it determines the weight of the briquettes. In addition, when briquettes differing in bulk density undergo combustion, briquettes having higher bulk density are able to retain their shape and generate more heat due to their better calorific values. Higher bulk densities also increases burning time of the briquettes. They as well depict better strength for storage and transportation. Briquettes with clay binder had higher bulk densities followed by briquettes with molasses (Figure 12). Large amount of clay used however compromised on the briquette quality because it lowered the calorific value. About 15% clay is acceptable.

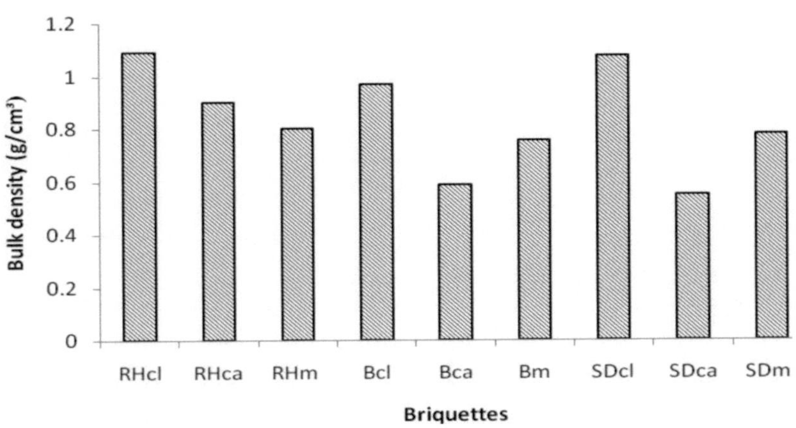

Figure 12. Bulk densities of briquettes. RH_{cl} – rice husk with clay binder, RH_{ca} – rice husk with cassava binder, RH_m – rice husk with molasses binder, B_{cl} – bagasse with clay binder, B_{ca} – bagasse with cassava binder, B_m – bagasse with molasses as a binder, SD_{cl} – saw dust with clay binder, SD_{ca} – saw dust with cassava binder, SD_m – saw dust with molasses.

CONCLUSION

Rice husk when compared to other biomass material used has a higher ash (20-25%) and potash (3.84%) content. Rice husk however, is an exceptional biomass because it has good flow ability, low moisture content (7-10%) and the ash contain fewer alkaline minerals. It makes an excellent fuel although its calorific value is less than those for sawdust and bagasse. The results of this work indicate that briquettes produced from bagasse and saw dust had more positive attributes than briquettes from rice husk. These attributes also agree with the proximate analysis carried out for the raw biomass materials.

REFERENCES

ASTM E711 –87. (2004) Test Method for Gross Calorific Value of Refine-Derived Fuel by the Bomb Calorimeter, ASTM International West Conshohocken

El-Saeidy EA. (2004). Technological fundamentals of briquetting cotton stalks as a biofuel. Ph.D Thesis, Faculty Agriculture and Horticulture, Humboldt University, Germany

EN 1860-2:2005) Test methods for charcoal and charcoal briquettes for use in barbecue appliances, CEN /TC 281, United Kingdom.

Enweremadu CC, Ojediran JO, Oladeji JT, Afolabi IO. (2004). Evaluation of Energy Potential of Husks from Soy-beans and Cowpea". Science Focus, 8:18-23.

GoK. (2007). Kenya Vision 2030. A Globally Competitive and Prosperous Kenya. Government of the Republic of Kenya, 2007. Ministry of Planning and National Development, and the National Economic and Social Council (NESC), Office of the President.

Ghorpade SS, Moule AP. (2006). Performance Evaluation of Deoiled Cashew Shell Waste for Fuel Properties in Bri-quetted Form. B.Tech. Thesis (unpub.), Dapoli, 15.

Grover and Tabil. (1997). Biomass briquetting technology and practices. Regional Wood Energy Development Programme in Asia GCP/ RAS/154/Net Field Document No.46.

Kaliyan N, Morey RV. (2009). Factors affecting strength and durability of densified biomass products. Biomass and Bioenergy 33:337-359.

Musa NA. (2007). Comparative Fuel Characterization of Rice Husk and Groundnut Shell Briquettes. *NJREDI*. 6(4):23-27.

Oladeji JT. (2011)."The effects of some processing parameters on physical and combustion characteristics of corncob Briquettes" Ph.D Thesis, Department of Mechanical Engineering, Ladoke Akintola University of Technology, Nigeria.

Olorunnisola AO. (2007) "Production of Fuel Briquettes from Waste paper and Coconut Husk mixtures" Agricultural Engineering International: The CIGR E-Journal Manuscript EE 06 006 pp123-128.

Wilaipon, P. (2007). Physical characteristics of maize cob briquette under moderate die pressure. American Journal of Applied Science, 4(12):995 – 998.

CALVIN ONYANGO obtained his BSc (1994) and MSc (1999) in Food Science and Technology from the University of Nairobi, Nairobi, Kenya. He obtained his PhD (2005) in Food Technology from Technische Universität Dresden, Germany. In 2008 he received a research award from Alexander von Humboldt Foundation to undertake postdoctorate research (2008-2010) at Max Rubner-Institut, Detmold, Germany. He is currently a research scientist in cereal science and technology at Kenya Industrial Research and Development Institute (KIRDI).

GÜNTER UNBEHEND obtained his Dipl. Ing. (1989) in Food Technology from Hochschule Ostwestfalen-Lippe, Germany. He has several years experience in the private sector as a food engineer. He joined Max Rubner-Institut in 1995. He currently heads the bakery department in Max Rubner-Institut.

MEINOLF G. LINDHAUER obtained his PhD in Biology from Westfaelische-Wilhelms-University, Muenster, Germany. He joined Max Rubner-Institut in 1992. He has 10 years experience in agricultural research and 18 years experience in postharvest cereal research. His experience in cereal research covers cereal chemistry and technology along the entire value-chain from raw material quality assessment through processing to end-product quality and nutrition. Prof. Dr. Lindhauer is a member of several international cereal research organizations such as AACC and ICC. In cooperation with the Association of Cereal Research e.V. Detmold, he is co-organizer of several international conferences (such as International Starch Convention) that take place every year in Max Rubner-Institut. He currently heads the Department of Safety and Quality of Cereals and is also the deputy President of Max Rubner-Institut.